William R. Park
CONSTRUCTION BIDDING FOR PROFIT

J. Stewart Stein
CONSTRUCTION GLOSSARY: AN ENCYCLOPEDIA REFERENCE
AND MANUAL

James E. Clyde
CONSTRUCTION INSPECTION: A FIELD GUIDE TO PRACTICE

Harold J. Rosen and Philip M. Bennett
CONSTRUCTION MATERIALS EVALUATION AND SELECTION:
A SYSTEMATIC APPROACH

CONSTRUCTION
BIDDING
FOR PROFIT

OTHER BOOKS BY WILLIAM R. PARK

The Strategy of Contracting for Profit. Prentice-Hall, 1966.

Cost Engineering Analysis: A Guide to the Economic Evaluation of Engineering Projects. John Wiley & Sons, 1973.

With James E. Kelly:

The Road Builders. Addison-Wesley, 1973.

The Airport Builders. Addison-Wesley, 1973.

The Tunnel Builders. Addison-Wesley, 1977.

The Dam Builders. Addison-Wesley, 1978.

With Sue C. Park:

How to Succeed in Your Own Business. John Wiley & Sons, 1978.

CONSTRUCTION BIDDING FOR PROFIT

WILLIAM R. PARK, P.E.
Consulting Engineer-Economist

A Wiley-Interscience Publication

JOHN WILEY & SONS
New York   Chichester   Brisbane   Toronto

Copyright © 1979 by John Wiley & Sons, Inc.

Library of Congress Cataloging in Publication Data:

Park, William R
  Construction bidding for profit.

  (Wiley series of practical construction guides.)
  "A Wiley-Interscience publication."
  Bibliography: p.
  Includes index.
  1. Building—Contracts and specifications—United
  States.   2. Contracts, Letting of—United States.
  I. Title.
  TH425.P29   658.1′5969   79-11451
  ISBN 0-471-04104-1

Printed in the United States of America

10 9 8 7 6 5 4 3 2

Series Preface

The construction industry in the United States and other advanced nations continues to grow at a phenomenal rate. In the United States alone construction exceeds two hundred billion dollars a year. With the population explosion and continued demand for new building of all kinds, the need will be for more professional practioners.

In the past, before science and technology seriously affected the concepts, approaches, methods, and financing of structures, most practitioners developed their know-how by direct experience in the field. Now that the construction industry has become more complex there is a clear need for a more professional approach to new tools for learning and practice.

This series is intended to provide the construction practitioner with up-to-date guides which cover theory, design, and practice to help him approach his problems with more confidence. These books should be useful to all people working in construction: engineers, architects, specification experts, materials and equipment manufacturers, project superintendents, and all who contribute to the construction or engineering firm's success.

Although these books will offer a fuller explanation of the practical problems which face the construction industry, they will also serve the professional educator and student.

M. D. MORRIS, P. E.

Preface

The contract construction industry is characterized by intense competition, low profit margin, and high failure rate. It accounts for a little less than 10 percent of the nation's gross national product, but for more than 17 percent of all business failures. The typical construction contractor operates with a 2 percent profit margin, earning only 10 percent on his invested capital.

Many of the construction industry's economic problems can be attributed directly to the ever-present cutthroat competition created by uninformed and unqualified contractors, which sets the unfair and unreasonable price levels under which the entire industry must operate.

It is difficult for a contractor to make a profit in competitive bidding when the highest price he can get for a job is the lowest price for which his cheapest competitor is willing to take the project.

Nevertheless, with adequate financial controls, sound management techniques, and an effective competitive bidding strategy, a competent contractor should earn a minimum of 4 to 5 percent on his total sales and 20 to 25 percent on his capital investment. The fact that less than one in five contractors is able to achieve even these modest levels of profitability offers ample evidence that profits of this magnitude do not just happen; they must be planned.

A profitable contracting operation requires a combination of technical knowhow, management competence, and business strategy. A contractor's profit is directly determined by three factors: cost, price, and volume. Each of these three factors, in turn, is influenced by other considerations that are largely beyond his control. Although he cannot control them, a contractor can—and must—react to them to maintain a profitable business.

While the contractor's labor, materials, and equipment costs continue to rise in an inflationary spiral, his prices are driven down by competitive pressures. Volume, his only apparent hope for financial survival, can be increased only by cutting prices further. However, profit—not volume—should be his primary objective.

The contractor must make full use of every strategic advantage that his technical, financial, and information resources can give him in order to operate profitably under the prevailing economic pressures. This can be accomplished by setting realistic goals, then employing all available information in determining the optimum combination of price, cost, and volume that will enable him to achieve his profit objectives.

Construction Bidding for Profit provides such information.

The author's earlier book on competitive bidding, *The Strategy of Contracting for Profit* (Prentice-Hall, 1966, now out of print) discussed the application of probability and statistics to practical problems. The same material, updated where necessary, is included in this book along with the abundant new material that has since become available. While the basic concepts have not changed since 1966, more effective ways of organizing and applying the information have been developed.

Information is a useful management tool only when it is available in a form that can conveniently be used as a basis for decision-making. Most contractors fail to make full use of the abundant competitive data available to them, largely because the data are not organized in a manner that reveals the valuable—and profitable—strategic insights into their competitors' behavior.

Recognizing that management's decisions can be no better than the information on which they are based and that competitive information is potentially the most valuable to the contractor, *Construction Bidding for Profit* brings together, in a convenient and usable form, all the information needed for construction management to make informed strategic decisions in competitive bidding.

The only additional data needed for developing an effective competitive bidding strategy in conjunction with the material presented in this book can usually be found in the contractor's own files of past jobs over the preceding year or two.

A major consideration throughout *Construction Bidding for Profit* has been to provide for a minimum of "hardware" and a maximum of "brainware." The whole approach is intended as a supplement to—never a substitute for—management judgment and experience.

The real key to success in competitive bidding lies in the thoughtful analysis of competitive information to formulate an effective strategy for use in a company's day-to-day operations. After the initial analysis has been performed and the system implemented and tested, it can easily be maintained and updated.

These bidding techniques have been successfully employed by contractors whose annual volumes range from less than $1 million to more than $200 million. The approach has proven equally effective whether the contractor's specialty is heavy construction projects involving millions of dollars per job, or subcontract work averaging only a few thousand dollars per job.

As long as competitive bidding is involved, the techniques presented in *Construction Bidding for Profit* can greatly increase the contractor's chances of realizing the maximum profits obtainable on his work. The specific benefits to be expected from applying the approaches described here will depend on the firm's operating characteristics and competitive situation. Whether a firm can benefit from using these techniques and how much additional profit can be generated from their use depends primarily on how successfully the firm is already operating.

The results of putting these bidding techniques into practice can be predicted only in general terms, based on the experience of other contractors. A rough estimate of what can realistically be achieved by incorporating these techniques in an existing operation can be obtained in the following manner:

1. Obtain the bid tabulations on about the last twenty jobs bid.
2. Figure the total amount of gross profits actually made on the jobs received—the difference between the successful bid prices and the estimated direct job costs on the work actually received.
3. Calculate the maximum gross profit that could have been made on the 20 jobs, had all competitors' bids been known in advance—the difference between the lowest competitors' bids and the estimated job costs.
4. Divide the gross profit actually made (from step 2) by the most that could possibly have been made (from step 3); this ratio or percentage represents the firm's bidding efficiency.

Most contractors are able to achieve bidding efficiencies only in the 20 to 30 percent range. Contractors applying the bidding techniques described in this book, though, have generally been able to achieve bidding efficiencies in the 40 to 60 percent range, about double their profits prior to adopting the scientific approach to competitive bidding. In other words, these contractors are now realizing profits that are about one-half as much as they could if they knew all their competitors' bids in advance.

The practical approach to contract pricing that is described here has been proven in practice by contractors in many different fields of construction. Most of the examples that are cited are drawn from the actual experience of these contractors. The techniques are not sophisticated from a mathematical standpoint, and no attempt has been made to make the facts conform to any preconceived theories. It is hoped that whatever is lacking in the way of theoretical precision is made up for in the ease of practical application. Any simple business strategy that actually works is rare and should be treasured by management.

The facts as presented herein are not always pleasant. But pleasant or not, they should be understood, related, interpreted, and acted on. They should never be ignored.

A thorough underderstanding of the problem and proper interpretation of the facts will enable the contractor to develop and carry out a strategic plan for achieving the highest level of profits possible under existing conditions. That is what business is all about. And that is what *Construction Bidding for Profit* is all about.

WILLIAM R. PARK, P. E.

Kansas City, Missouri
March 1979

Contents

Tables

Illustrations

COMPETITIVE BIDDING (SIMPLIFIED)

How is a contractor chosen to build a road? Many contractors may wish to build the road.

Each contractor writes on a piece of paper the amount of money he would charge for doing the job. This amount is called a *bid*. The contractor puts his bid into an envelope and drops it into a box with all of the bids from the other contractors.

When the time comes to look at the bids, everyone is very excited, because the contractor whose bid is the lowest will get the job.

Who will it be?

from *The Roadbuilders*
a book for beginning readers
(kindergarten through third grade).

1

What to Expect from Strategic Bidding

Expect a Miracle.

ORAL ROBERTS

We don't just expect miracles;
we depend on them.

ANONYMOUS CONTRACTOR

To be successful in his competitive bidding strategy, a contractor needs to bid high enough to assure himself of a profit on each job, yet low enough to get the job. His problem is that if he bids high enough for a sure profit, he is too high to get the job. But the only way to be sure of having the low bid is to bid below cost.

Under these conditions, it seems that a contractor is then faced with two extremely unpleasant alternatives: (1) an excellent chance of making no profit with a low bid, or (2) no chance at all of making a high profit with a high bid. However, somewhere between these extremes there is an opportunity for a contractor to make a reasonable profit.

Every job has an optimum or "best bid" which will, in the long run, result in the highest possible profit obtainable under the existing competitive situation. The objective of a competitive bidding strategy is simply to identify this optimum markup for a job before it is bid, or better yet, before incurring the time and expense of preparing a detailed cost estimate.

A "Fair Profit"

What constitutes a "fair" profit for a construction contractor? A typical general contractor should earn a *minimum* of 4 to 5 percent on his total sales volume and a return on his net capital investment of 20 to 25 percent.

1

By applying just a few strategic principles to his bidding practices, *he can make about half as much profit as he could if he knew all his competitors' bids in advance!*

If this much profit still fails to bring him his 4 or 5 percent profit margin and his 20 to 25 percent return on investment, then he is either (1) in the wrong business or (2) going after the wrong jobs.

Measuring Bidding Efficiency

In contract construction, where the highest price that a contractor can get for his work is the lowest price that his cheapest competitor is willing to take to do the job, efficient bidding is obviously the key to success.

Bidding efficiency is defined simply as the ratio of the amount of profit actually made, to the amount that *could* have been made had all competitors' bids been known prior to the letting. In other words, it represents the amount that could have been made by taking all jobs at the lowest competitors' price, providing the jobs would have been wanted at those prices.

Table 1 summarizes the actual results achieved by a contractor on nine recent jobs, of which the contractor was low bidder on four. These four jobs had estimated direct costs totaling $302,800, and the profit realized on them was $26,700, assuming the cost estimates were correct.

The markups applied to these nine jobs ranged from 5 to 14 percent, based on the contractor's "feel" for the competitive situation at the time. In each case, the lowest competitor's bid on a job determines the maximum profit that can be

TABLE 1. Summary of Bid Tabulations on Nine Contracts

Job Number	Number of Competitors	Lowest Competitor's Bid	Estimated Cost	Actual Bid	Maximum Profit Potential	Actual Profit
1	2	$ 95,400	$ 87,600	$ 96,300	$ 7,800	$ 0
2	5	41,400	36,300	40,600	5,100	4,300
3	4	416,500	388,300	428,900	23,200	0
4	2	642,200	633,100	665,300	9,100	0
5	6	127,900	112,200	122,200	15,700	10,000
6	3	15,600	14,700	16,600	900	0
7	1	26,800	22,400	25,500	4,400	3,100
8	3	146,400	131,900	141,200	14,500	9,300
9	5	478,900	457,500	498,500	21,400	0
Total			$1,884,000		$107,100	$26,700

made on that job, so the "maximum profit potential" represents the difference between the lowest competitor's bid and the estimated direct project cost.

Therefore, the maximum profit potential defines the most profit that could possibly have been made even if all competitors' bids had been known in advance; for these nine jobs, then, the maximum profit potential was $107,100.

The $26,700 profit actually realized by this contractor on the four jobs he got represents 24.9 percent of his maximum profit potential on the nine jobs:

$$\frac{26,700}{107,100} = 0.249 = 24.9 \text{ percent}$$

A bidding efficiency in the 20 to 30 percent range is typical of many contractors who operate intuitively "by the seat of their pants."

Effect of Different Markups

Table 2 shows what the effect on this contractor's profits would have been had different markups been applied on the same nine jobs described in Table 1. This is an extremely useful exercise for any contractor, and will frequently reveal things that he might never have imagined.

Table 2 shows that the alternatives range from getting all nine of the jobs at a 0 percent markup (no profit) to getting none of the jobs at a 20 percent markup (same result, no profit). Between these extremes, the choices will produce varying levels of profits ranging from $3400 at a 15 percent markup to $54,500 at the 7 percent level.

As the markup increases from zero, the volume of work decreases. Eventually, profits increase to a maximum, then again decline to zero at the point of no volume. Figure 1 shows these relationships. Had all jobs been bid at a 7 percent markup, this contractor's bidding efficiency would have been

$$\frac{54,500}{107,100} = 0.509 = 50.9 \text{ percent}$$

which is very good.

Competitive Strategy

By analyzing the bids on many recent jobs—about 50 in this case, but preferably 100 or more—a competitive bidding strategy was developed for this contractor as shown in Table 3. This bidding strategy identifies the markup to

TABLE 2.   **Results of Applying Various Percentage Markups to Bids**

Markup Applied to Job (%)	Number of Jobs Won	Estimated Cost of Jobs Won	Gross Profit on Jobs Won	Bidding Efficiency (%)
0	9	$1,884,000	$ 0	0
1	9	1,884,000	18,800	17.5
2	8	1,250,900	25,000	23.3
3	8	1,250,900	37,500	35.0
4	8	1,250,900	50,100	46.7
5	7	793,400	39,600	37.0
6	7	793,400	47,600	44.4
7	6	778,700	54,500	50.9
8	5	390,400	31,300	29.2
10	4	302,800	30,200	28.2
12	3	170,900	20,600	19.2
15	1	22,400	3,400	3.2
20	0	0	0	0

be used for different combinations of job sizes and competitive conditions. It indicates, for example, that a job estimated at more than $200,000 and bid against one or two competitors should carry an 11 percent markup.

Were more data available for analysis, say several hundred jobs, the bidding strategy could be broken down in more detail. This particular strategy varies the markup from a low of 4 percent on large jobs bid against many competitors, to a high of 21 percent on small jobs bid against few competitors.

The results of applying this particular bidding strategy to these same nine jobs is shown in Table 4. Total profits of $61,800 would have been realized on five jobs costing $1,126,200. This profit is 13 percent more than would have been possible using any fixed markup, and is 130 percent more than the contractor actually made by bidding intuitively. This bidding strategy raises the

TABLE 3.   **Optimum Percentage Markups for Different Job Characteristics**

Number of Competitors	Under $50,000	$50,000 to $200,000	Over $200,000
1 to 2	21	15	11
3 to 4	11	8	6
5 to 6	8	6	4

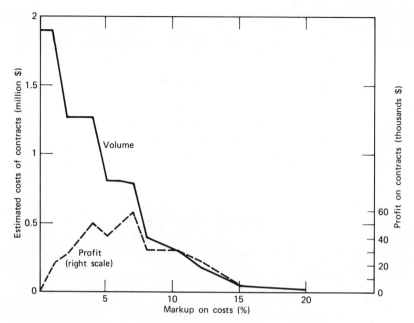

FIGURE 1. Profit and volume associated with different markups.

contractor's bidding efficiency to 57.7 percent, about as high as any contractor could reasonably hope for.

In this case, the strategy was developed from data on about 50 jobs *other than the 9 jobs on which it was tested.* Any strategy developed on the basis of historical data will look good if tested on the same jobs that it was developed

TABLE 4. Results of Using Optimum Markups

Job Number	Number of Competitors	Estimated Cost	Optimum Markup (%)	Optimum Bid	Profit Using Optimum Bid
1	2	$ 87,600	15	$100,700	$ 0
2	5	36,300	8	39,200	2,900
3	4	388,300	6	411,600	23,300
4	2	633,100	11	702,700	0
5	6	112,200	6	118,900	6,700
6	3	14,700	11	16,300	0
7	1	22,400	21	27,100	0
8	3	131,900	8	142,500	10,600
9	5	457,500	4	475,800	18,300
Total					$61,800

from. Such a test obviously proves nothing but good hindsight, which all contractors already have.

The Real Payoff

The only true test of any bidding strategy is in practice. If the contractor can achieve higher profits with it than without it, then it is a good strategy (or at least a better one than he had before). Otherwise, it is worthless except for its educational value; often, a great deal can be learned just by going through the analysis.

But there comes a time when a person should stop paying to acquire an education and start collecting for having one. Given a choice, profit is preferred to an education, and the contractor should set his sights in the direction of increased bidding efficiency. This will ultimately be reflected in better jobs and higher profits; not necessarily *more* jobs or *bigger* jobs, but jobs offering better opportunities for profit.

Benefits of Strategic Bidding

The results of applying a carefully thought out bidding strategy can be quite impressive. How much improvement can be made depends on how well the contractor is doing prior to adopting the strategy. Most contractors have bidding efficiencies in the 20 to 30 percent range in their normal operations.

The same contractors, with a bidding strategy tailored to their specific operations, can usually achieve bidding efficiencies in the 40 to 60 percent range, about double their previous results.

This means they can make twice as much profit as before on the same types of jobs!

Table 5 shows the results achieved by several different contractors whose bidding efficiencies rose to the 42 to 56 percent range from their previous 20 to 30 percent levels just by applying appropriate bidding strategies.

TABLE 5. Results of Strategic Bidding for Different Types of Contractors

Type of Contractor	Number of Jobs Bid	Number of Jobs Won	Maximum Profit Potential	Actual Gross Profit	Bidding Efficiency (%)
General (heavy)	36	16	$1,808,000	$1,014,000	55.8
General (buildings)	32	16	151,700	68,400	45.1
Sub (building trade)	107	44	386,600	162,200	42.0

In addition to providing overall pricing guidelines, a bidding strategy will enable the contractor to do the following:

- Determine the chances of getting a job by bidding with any given markup.
- Identify the markup that will result in the greatest possible profit on a specific job in view of the prevailing competitive situation surrounding that job.
- Select from a number of different projects, the jobs offering the greatest profit potential.
- Decide whether a particular job offers sufficient profit potential to justify submitting a bid at all.

Even if a bidding strategy does nothing more than sharpen the contractor's intuitive sense by providing him with objective information about his competitors, the time and effort spent in its development will have been generously rewarded.

2

An Introduction to Competitive Business Strategy

The task of management is not to apply a formula but to decide issues on a case-by-case basis. No fixed, inflexible rule can ever be substituted for the exercise of sound business judgment in the decision-making process.

ALFRED P. SLOAN, JR.

Construction management's most difficult and most persistent problem, as well as its greatest challenge, is the "profit squeeze," which has been tightening since the prosperous decade following World War II. While the gross volume of construction activity has continued to increase, the net profits resulting from the work have continued to shrink. With little immediate likelihood of any major reversals in this unpleasant trend, management must continually be alert for more effective ways of improving the profit picture. Intuition and guesswork alone, perhaps adequate as the basis for management decisions during highly prosperous times, are no longer sufficient to meet the pressures of increased competition.

The Role of Strategy

Strategy involves making competitive decisions; the more competitive an industry, the greater the need for making sound strategic decisions. Nowhere is this need greater than in construction contracting, where competitive pressures are probably more intense than in any other industry.

8

Competitive bidding, especially, offers abundant opportunities for the application of strategy. In competitive bidding the contractor is faced with two seemingly incompatible and contradictory objectives: He must bid high enough to make a profit, yet low enough to get a job—*both at the same time!*

Most contractors are able to estimate, at least with reasonable accuracy, what their direct costs on a job will be. And taking a single job, most of the competing contractors can also be expected to encounter roughly the same costs of performing the work; they are all subject to the same costs of operation, have access to the same labor supply, use the same types of equipment, obtain supplies and materials from the same sources, and have somewhat comparable, if not equal, supervisory capabilities.

Still, some contractors are able to operate successful, profit-making businesses, while others go broke or are, at best, barely able to survive. This is in spite of the fact that most are technically competent and are capable of actually performing the required quality of work.

What, then, makes the difference between success and failure in the contracting business? Most of the difference can be found in management's ability to make the strategic decision at the strategic time. Strategy is an essential ingredient for success in every competitive situation.

The strategy of successful bidding depends more on management's judgment than on any other single factor. It is a proven fact that more business failures are caused by poor judgment than by a lack of knowledge or technical competency. And despite the cries of unsuccessful businessmen, the causes of business failure can seldom be attributed to bad luck—just to bad business practices and poor business strategy.

The Meaning of Strategy

Strategy can be defined in the following ways:

- Skillful management in getting the better of an adversary.
- The means by which a company uses its financial and physical resources to accomplish its objectives.
- The science and art of meeting competition under the most advantageous conditions possible.
- A careful plan or method.
- The art of devising or employing plans toward a goal.
- Management's ideas regarding a firm's objectives, the means by which these objectives will be accomplished, and the reasons for pursuing them.

In all cases, strategy involves management's decisions in the face of uncertain competition. A strategy is primarily concerned with attempting to meet this uncertain competition under favorable terms, ideally, in matching one's major strengths against the competitors' major weaknesses. Games which combine elements of both skill and chance are good examples of the application and importance of strategy.

Poker, in many respects, bears a close resemblance to business in general and to the contracting business in particular. In poker everyone has access to the same hands over the long run; everyone must play under the same established set of rules; and the player's resources, if not equal, are at least similar. Still, there can be found the consistent winners, the steady losers, and the many in-between who neither win much nor lose much but who nevertheless stay in the game just because they like to play.

The similarity between poker and the contracting business is especially striking in several respects:

1.   There is an element of chance involved, but a much greater element of skill and judgment. In the long run the big winners will be those who most effectively exercise this skill and judgment.

2.   The skill in the game is based on judgments of the probability of different events occurring, and a sound and thorough knowledge of the chances of these events happening is essential for success. These probabilities can be calculated mathematically but are usually recognized and applied instinctively or intuitively.

3.   The relative competitive strength of a hand (or bid) depends to a large extent on the number of players (or bidders) in the game. What would probably represent a winning hand in a 2-handed game would be a frequent loser in a 5-handed game and might be completely worthless in a 10-handed game.

Success in business, like success in poker, requires certain general attributes on the part of management: a thorough understanding of the rules and conditions of the game, a sound knowledge of the odds of the game, and a good sense for strategy.

The contractor's overall strategy in competitive bidding is to secure the combination of elements that will, in the long run, enable him to achieve certain desired objectives. Almost any sales volume can be generated by a contractor if his bids are placed low enough, even though the jobs that he wins are unlikely to result in a profit. Conversely, his markup can be raised to any level he desires, but with the ultimate result that his sales volume will be reduced to so low a point that there will be no profit. Somewhere between these unpleasant extremes of a high markup and no volume, or a huge profitless volume with no markup, the contractor must find a markup which is compatible with his firm's objectives and which will allow him the maximum profit possible under the existing competitive conditions.

The Profit-Making Area

In contracting the area where profits must be made is the area between the estimated direct cost of doing a job and the amount bid for the job. The lower limit of a job's price—the actual, out-of-pocket direct cost of the work—is more or less fixed, since there are few contractors who will willingly (or at least knowingly) take a job at less than their direct cost. The upper limit of a job's price, on the other hand, may be established by several important factors including the following:

1.   What the contractor considers a fair return on his invested time and capital, based on his own appraisal of the risks associated with the particular job and on what he could earn elsewhere with his capital.

2.   How badly he needs the work to keep his men and machines occupied, thereby providing a larger base over which overhead and indirect costs can be spread.

3.   How much the traffic will bear—how much of a markup he thinks he can add to the direct job costs and still have some chance of getting the job.

The first two of these factors depend primarily on the individual contractor's own business operation; the third factor depends chiefly on the actions of his competitors, which he can anticipate but cannot control.

The profit-making area between the estimated direct cost and the bid price is also the area where most losses are incurred, either through management's failure to include sufficient markup or through its failure to attain sufficient volume to recover indirect costs. And it is therefore within this area that management is offered an opportunity to prove its worth, for here the opportunities for managerial discretion and control are presented, and, in fact, cannot be avoided. Probably the most important single responsibility of construction management is to make decisions within this area, and the success or failure of nearly every construction firm can be traced directly to these management decisions.

The Management Decision

The major function of any executive is to make decisions. In general, the more difficult the decision, the greater the rewards for making the correct, or at least the best possible, decision. In order for the executive to be able to make effective decisions, he must first have alternatives to choose from, information (both factual and theoretical) on which to base his decision, and, finally, the authority to act on his decisions once he has made them.

Every important management decision involves a process of selection between two or more alternative courses of action; otherwise, no decision would be necessary. In making the best possible decision under the existing, known circumstances, several well-defined steps should be followed:

1. Define the objectives; what is hoped to be accomplished by the decision?
2. Identify different approaches that might be used to attain the desired objectives.
3. Analyze and compare these alternatives to see which offer the best possibilities.
4. Choose the approach that is best suited to the firm's capabilities and objectives.

The first step in developing a suitable competitive strategy is to determine the firm's objectives. These objectives can be expressed and measured in a number of different ways. Profit, probably the most common business objective, can be measured, for example, as a net dollar amount, or as a percentage return on sales, investment, net worth, assets, or working capital. Other objectives, such as gross sales volume or a specified share-of-market percentage, might also be used.

A single objective can frequently be achieved in several different ways. For example, a profit objective might be attained by means of a low markup on a large volume of work; this approach, however, might fail to meet a minimum return-on-invested-capital requirement. At the other extreme, a low volume of work taken with a high markup might yield a high rate of return on capital but fail to capture the desired share of the total market.

By carefully and objectively analyzing the possible outcome of each of the alternative means of reaching the desired objectives in the light of existing and expected competitive conditions, the effect of each course of action on the company can be estimated. Can the large volume of work necessary with a low markup be achieved, even with the low markup, in view of the policies and pricing tactics of competitors? Or will it be more feasible to obtain the relatively small volume of work required to meet profit objectives with a higher markup?

Usually, the best answer will lie somewhere between the two extremes. The important thing for the executive to recognize is that, for every competitive situation, there is *some* decision that will over the long haul enable him to be more successful, as measured by his own standards or definitions of success, than will any other.

The Strategic Decision

Even the most common types of strategic decisions that face construction management, such as the pricing of a bid, involve an immense number of variable factors. A computer would be hard-pressed to process all of the factors, if they could all be identified (and they cannot be). The human mind is hopelessly inadequate to cope with even a small part of the available information. The executive who, through insights gained through his background of experience, skill, and judgment, can arrive at even a moderately satisfactory decision on the basis of a brief examination and superficial analysis of incomplete data will probably fare much better than will most of his competitors.

The most difficult types of management decisions that must be made are those which involve a high degree of uncertainty, especially when the uncertainty is in regard to the possible actions of competitors. These are strategic decisions and often must be made solely on the basis of management's informed judgment.

Decision Theory

Formal decision theory, as such, encompasses a wide variety of analytical techniques for handling management problems. The purpose of decision theory is to solve, through objective methods, management problems of the type that traditionally could be attacked only through intuitive judgment. Decision theory, as a formal discipline, has evolved almost entirely since the beginning of World War II.

Decision theory includes a fairly well-defined assortment of analytical techniques for solving many different types of complex business problems, including the common problem of keeping up with the immense number of decisions that must be made by management.

An effective way of increasing management's capacity to manage is to eliminate the need for top-level decisions on routine matters. Many of the problems faced by construction management are of a routine nature, and the decisions relating to them can therefore be handled in a routine manner, requiring little effort on the part of management. Policies can be formulated regarding the action and procedures to be followed in routine matters, and responsibility for carrying them out can be easily delegated. Since the facts on which routine decisions are based are usually clear-cut, there is little risk involved in their outcome and little need for top management to be overly concerned. By thus relieving top management of the necessity for making routine decisions, the executive can be freed to devote more of his effort to the more difficult and frequently the more critical and rewarding decisions.

Many of the decisions that management is called on to make must be made without the benefit of factual data. These are truly the decisions that management is paid for making—the decisions that others simply are not capable of making. These decisions require the executive to apply all of his mental resources—intelligence, experience, imagination, and intuition—in the face of uncertainty. The tools of decision theory can help lessen the dangers of certainty.

Decision-Making Tools

Certain decision-making tools have been developed that can remove some of the burden from management in making strategic decisions. These decision-making aids can provide a valuable supplement to the executive's personal skill and experience, although they can hardly be expected to substitute for his informed judgment. Some of the most powerful decision-making tools for construction management include the following:

1. Statistics.
2. Probability theory.
3. Operations research.
4. Game theory.

Statistics

Statistics deals with the collection, tabulation, analysis, presentation, and interpretation of factual data. Statistics can be advantageously employed whenever large numbers of facts need to be clarified before the data can be used as a basis for drawing useful conclusions. A knowledge of statistics is valuable to management in planning, controlling, and organizing activites; a familiarity with statistical techniques can enable the executive to use his experience most effectively. The manager with a broad background of experience can undoubtedly forecast his own company's sales picture and business outlook better than even the most capable statistician, but a forecast based on a thorough analysis of the best data available *plus* a background of managerial experience will be much better than a decision based only on experience, intuition, and hunches.

Probability Theory

Probability theory is a separate branch of statistics which deals with methods for determining the likelihood of occurrence of a particular event. Probability theory began, logically enough, in connection with gambling. However, business is frequently as great a gamble as is gambling itself, and the application of probability theory to business problems is a natural development. Probability theory relies both on theory and experience for its foundation. Past experience may often be the best guide to future actions in many business activities, while others may lend themselves to a more theoretical or mathematical treatment. In any event, most important business decisions involve judgments of probabilities concerning uncertain future events. When it is not within management's power to ascertain what is actually going to happen in a competitive situation, strategic decisions will necessarily be based on management's appraisal of what is most likely to happen. The executive, in effect, must somehow link the known past with the unknown future. When in doubt regarding the future, decisions should be made according to the best possible judgment of the odds. Determining the odds in an objective manner is the main purpose of probability theory.

Operations Research

Operations research is the application of research methods (the scientific method) to problems involving the functioning of a business organization, with the ultimate objective of increasng the organization's efficiency and effectiveness. In general, the procedure followed in most operations research studies is to use some sort of mathematical model to represent the problem so that alternative solutions can be devised, tested, evaluated, revised, and retested before a solution is finally put to work. This procedure greatly minimizes the risk of bad decisions or policies, and allows for more alternatives to be studied in a limited time. Five basic types of problems lend themselves most readily to the techniques of operations research:

1. Inventory problems, involving the size of inventory to be carried and the timing of purchases.
2. Allocation problems, where limited resources must be spread among alternative uses.

3.   Waiting-time problems, where jobs must be planned and scheduled to minimize down-time and cost and to increase equipment and manpower utilization.

4.   Repair-replacement problems, in determing the optimum level of equipment maintenance that is justified, and in deciding exactly at what point replacement becomes more economical than repair.

5.   Competitive problems, in which the actions of competitors must be predicted on the basis of their past performance, and the firm's strategic decisions must be based on these predictions.

Operations research, then, studies the past to determine facts, develops theories that attempt to explain the facts in mathematical terms, and uses these theories to predict future happenings and to examine the effects of alternative actions.

Game Theory

The theory of games covers the methodology for analyzing competitive situations involving two or more opposing interests. Each player is attempting to come out of the game with maximum gains, which can be accomplished only at the expense of the other players in the game. Game theory, therefore, attempts to define the general rules and procedures by which each player can plot a strategy that is best suited to his objectives. In arriving at an "optimum" strategy, each possible strategy must be matched against each of the opponents' possible counterstrategies, and the results of all the possible outcomes must be evaluated. Ideally, the optimum strategy will result in the best possible outcome for a player when he is confronted by his opponents' best strategies; if an opponent should slip and use something less than the best possible counterstrategy, however, the player using an optimum strategy will win even more, while the less-than-best strategist will end up with less than he could have attained otherwise. A general knowledge of the principles of game theory can be a great help to management in clarifying its understanding of complex competitive situations, and game theory techniques can form a solid framework with which many otherwise difficult competitive problems can be conveniently solved.

Using the Tools

These four related disciplines—statistics, probability theory, operations research, and the theory of games—provide the executive with an exceptionally useful and valuable set of tools for using his own skill and experience to maximum advantage in making difficult decisions and in developing appropriate strategic plans for his firm.

The company's strategic plan will necessarily be broken down into several parts perhaps including a long-range (several years) objective, a short-range (one year) objective, and an immediate, job-by-job objective. These objectives must, of course, be compatible with each other.

By using the analytical tools available to him with discretion and in conjunction with his own professional skill and judgment, the contractor can benefit in several important ways:

1. He will be able to objectively consider and evaluate a larger number of jobs in less time than would otherwise be possible, eliminating those jobs which involve excessive risk or competitive pressure or which do not fit in with the firm's objectives or capabilities.

2. He will be able to decide which jobs to bid based on their relative profitablility or desirability, thereby directing his efforts and his company's resources toward the most potentially favorable opportunities.

3. He will be able to determine what level of markup to apply to each job, in keeping with his firm's objectives and in light of the competitive conditions to be encountered on each.

4. Should he for any reason choose to digress from his theoretically optimum strategy, he will be able to determine objectively the effect of his digression.

There need be no perfect strategy. The main purpose of strategic planning for construction management is simply to enable the executive to select a course of action that best suits his firm's needs and purposes, that is possible to attain, and that strikes a good balance between the possible gains and the potential risks.

A competitive business strategy cannot insure a profit, nor can it offer a substitute for management experience, judgment, and skill. Business in general, and especially the contracting business, will always remain to a considerable extent a gamble. The contractor must remember that a good gambler always tries to get the best possible odds; in contracting, the odds improve when management acts on the basis of a careful analysis of the industry and the competition and on a thorough understanding of his own operations.

Future profits depend on present decisions. If present decisions are based on informed, intelligent management judgment, the success of the business is assured.

Summary

Intuition and guesswork are no longer adequate to insure profitable operations in the contracting business. A well-planned strategy leading to the best possible decisions in the face of uncertain competiton is required. An important part of the firm's strategy will be centered on the markup—the amount added to direct job costs—since it is within this area that profits must be made. An objective approach to strategic planning is necessary, and new analytical procedures are now available for aiding management in the decision-making function. Decision theory, one of the most valuable management aids, employs several important "tools," including statistics, probability theory, operations research, and game theory. These tools provide a valuable supplement to—but never a substitute for—informed management judgment in helping to simplify and solve a variety of complex strategic business problems.

3

The Strategic Approach

The strategic aim of a business is to earn a return on capital, and if in any particular case the return in the long run is not satisfactory, then the deficiency should be corrected or the activity abandoned for a more favorable one.

ALFRED P. SLOAN, JR.

Strategic planning for the contractor involves the development of realistic objectives, and the establishment of specific programs, procedures, and policies to achieve these objectives. The planning must include all of the firm's major activities, especially those activities that affect the three main determinants of a firm's profits: cost, price, and volume.

The term strategy implies a situation in which a number of alternatives are available, one of which is better than the others in acomplishing the firm's objectives. Given such a choice of alternatives, the strategic plan is simply a formalized or standardized procedure for finding the best possible alternative under the existing conditions.

Strategic situations confront the contractor every time he considers a new project. His first decision must be *whether* to bid on the job. Then, if competing for the job appears to be compatible with his objectives, he must decide *how much* to bid.

Given several available jobs on which he has an opportunity to bid, the contractor's strategic plan should tell him the following:

- Whether a job is worth bidding at all.
- In what sequence the various available jobs should be bid.
- How much to bid on each one.

A bid low enough to assure getting a job will invariably be too low to guarantee a profit. On the other hand, a bid high enough to assure an adequate profit margin usually has only a remote chance of winning the job. In extreme cases, the choices would appear to be between an excellent chance of making nothing or no chance at all of making a lot.

However, for every job there exists some pricing strategy that will result in the best possible combination of the chances of getting the job and amount to be made by getting it at that price. This is the *optimum price,* which in turn defines the *optimum markup*.

While even this optimum markup cannot guarantee a profit in the long run, it will assure the highest profit (or lowest loss) that is *possible* under the existing competitive conditions.

Benefits of Strategic Planning

A carefully conceived strategic plan will enable a company to respond to its environment, rather than be controlled by it. The firm can take advantage of new opportunities at the appropriate time, rather than reacting in a defensive manner to each new competitive situation.

Strategic planning will help the firm chart a desirable long-term course, even while more effectively meeting short-range problems. Many of the common day-to-day operating problems can be quickly resolved by reference to the strategic plan.

The competitive strategy will help identify in advance the optimum markup on each prospective job, considering the characteristics of both the job and the competition. A good competitive strategy will assist management in the following:

- Determining the chances of being low bidder on a job by applying a given markup.
- Identifying the markup that will result in the greatest possible profit on a specific job, in view of the competitive factors affecting that job.
- Selecting from a number of different projects the jobs offering the highest profit potential.

The strategic plan is set up so that it can be tested before it is actually applied. Thus, it allows the contractor to experiment with his ideas before he risks his money. The firm is able to investigate the likely consequences of many different courses of action without making any major financial commitments.

Major Elements in the Strategic Plan

There are four major elements in any strategic plan:

1. Objectives
2. Business plan.
3. Goals.
4. Monitoring and updating.

Objectives

Objectives are the broad statements that provide the primary guidelines for the development of the firm. Objectives may be to attain a specified percentage return on invested capital, to achieve the maximum possible gross profit on the work performed, or something else. Whatever they are, the objectives must be clearly defined.

Business Plan

The business plan consists of the tactical methods and procedures that will be used to achieve the predetermined objectives. The business plan provides day-to-day guidance for the firm's management in such tasks as screening prospective jobs and deciding how each job should be priced in view of the prevailing conditions.

Goals

Goals include the target dates, the assignments of responsibility, and the specific benchmarks that will be continually referred to in measuring the firm's progress toward its objectives.

Monitoring and Updating

How well the firm is accomplishing its objectives must be continually surveyed by reference to the planning goals. If progress is lagging, either the goals are unrealistic or the firm is not functioning as it should. In either case, corrections should be made before the problem exceeds the solution and the firm finds itself in real trouble.

The Steps Involved in Strategic Planning

The entire procedure for developing a strategic plan can be summarized in just nine steps, as follows:

1. Identify the problem.
2. Define the objectives.
3. Collect the data.
4. Interpret the data.
5. Devise alternative plans.
6. Test the alternatives.
7. Select the best plan.
8. Put the plan into practice.
9. Evaluate the results.

An often-used tenth step, "affix the blame," is not considered here.

Some of these steps can be pursued at the same time, with considerable overlapping from one to the next. Working on each subsequent step will, in fact, help to sharpen the understanding and enhance the usefulness of each preceding step.

The entire planning process may take anywhere from one month to one year, depending on how complex the operations are. Actually, the strategic plan can never be considered complete, since the results will be continually reviewed and reevaluated throughout its life.

Figure 2 shows a vague sequence chart for the strategic planning process. The most time-consuming steps involve data analysis and plan formulation— what might be considered the "creative" portions of the program.

1. *Identify the Problem.*   Most contractors are aware that many problems exist in the construction industry. Many of these problems are completely beyond the individual's control, and others would require an unprecedented degree of cooperation among contractors for their solution. In fact, most of the problems facing the contractor may not be solvable. But making profits, not solving problems, is the contractor's objective. He must therefore recognize and understand the problems that will influence his choice and pursuit of objectives. Basically, the industry's problems can be summarized as a situation where costs are rising faster than prices, and volume must make up the difference to keep total profits at a constant level. The individual contractor's problems can be briefly described as too many contractors competing too fiercely for too little work.

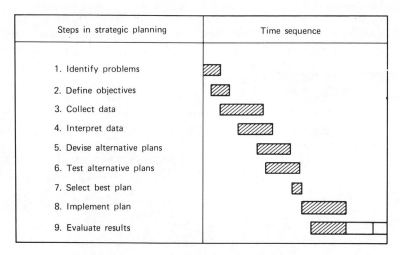

Steps in strategic planning	Time sequence
1. Identify problems	
2. Define objectives	
3. Collect data	
4. Interpret data	
5. Devise alternative plans	
6. Test alternative plans	
7. Select best plan	
8. Implement plan	
9. Evaluate results	

FIGURE 2. The steps involved in strategic planning.

There are several observations that should be mentioned here regarding the approach to problem-solving in general:

- Do not attempt to solve a problem without first making sure that there is one.
- If there is a problem, avoid putting a lot of effort into attempting to solve it unless there is some reason to believe that it *can* be solved.
- If there *is* a problem that *can* be solved, be certain that there is some benefit to be gained by its solution before spending much time working on it.
- Solve the important problems first. There are plenty of big problems to worry about without wasting time on little ones.

2. *Define the Objectives* Defining the firm's objectives can be a complex and difficult task because of the variety of possibly conflicting goals. The contractors's objective is necessarily a compromise between how much he would like to make, how much he needs to make, how much it is possible for him to make, and how much he should make. What he *should* make is a fair return (say 20 percent or more) on his invested capital. What he *needs* to make depends on his own operations and cost structure. What he *can* make depends on a combination of the total amount of work available and the pricing policies of his competitors. What he would *like* to make or *needs* to make have little relationship to the others, except that if he *needs* or *wants* to make more than

he *can*, he will obviously be less than pleased with his results. Setting realistic objectives requires that the *minimum acceptable* and *maximum possible* profits not have too great a gap. They must be defined for the firm in view of its own particular situation within the overall competitive environment.

3.   *Collect the Data.*   Data collection can begin almost immediately, even before the objectives have been finally decided on. All the data pertaining to the contractor's problems should be collected and maintained in current order. Since most of the major problems can be classified as referring either to the company, the competition, or the economic environment, this suggests the types of data that might prove most useful. *Costs* reflect the efficiency of a company's operations, and the contractor needs to know several things about his own costs:

- How accurate are his estimates? Final costs on all projects should be compared with the estimates. The easiest way in the world to get a job is to make a bad estimate.

- Are costs in line with those of other contractors in the same field? Does anyone have a particular advantage with respect to operating efficiency, or have access to cheaper labor, materials, or equipment?

- Can costs be reduced by applying new management techniques in purchasing, scheduling, financing, inventory control, equipment maintenance, and so on?

Competition sets the upper limit on prices. Typical competitive information that should be collected for every job on which the data can be obtained includes the following:

- The size and type of job.
- Any distinguishing or unusual characteristics.
- Estimated direct, out-of-pocket costs of performing the specified work.
- How many competitors were involved?
- How much did each one bid?
- How much did the low bidder leave "on the table?"
- How much profit could have been made, if any, by taking the job at the lowest competitor's price?

The *economic environment* determines the amount and type of work that has been and will be available. By keeping up-to-date on what is going on in the industry and the economy, the contractor will have some idea of what to expect in the future. If he has the right kind of economic data available, for example, a forecast of 10,000 new housing units to be built in his area the next year can be translated directly into meaningful terms—like dollars—for his own operations.

4. *Interpret the Data.* Interpretation of the data must necessarily wait until most of the data have been collected. The raw data should be organized in such a way that each piece of information can be found when it is needed. If the data are properly organized they can be analyzed without too much difficulty; some of the answers will, in fact, be immediately obvious. If the data show, for example, that the firm has never gotten a job that was bid against 10 or more competitors unless it made a mistake in its bid, then a reasonable interpretation of the data would be that it does not pay to bid jobs under these conditions. The whole objective of data organization and analysis is simply to help make some sense out ot all those numbers—specifically, to *interpret* them into meaningful terms that will help the contractor get work and make a profit. Analysis of the data is one of the most fascinating parts of the whole planning sequence. While it does not necessarily show the contractor where he can expect to go in the future, it at least tells him where he has been in the past and why. Then, if he does not want to go there again, he can attempt to avoid repeating the things that sent him there the first time. Experience, it is said, is what enables you to recognize a mistake when you make it again. Nevertheless, what has happened in the past usually provides a good indication of what the future holds. The data provide a sound and objective basis for gaining a knowledge of the past and an understanding of the present.

5. *Devise Alternative Plans.* Alternative plans can be proposed while the data interpretation phase is still in progress. As noted in the preceding section, future plans should be based on past experience unless there is an exceptionally good reason to do otherwise. Once the past and present are thoroughly understood, it is relatively easy to figure out how one might go about operating under the conditions imposed by the economic environment and the competition. Given the data to work with, along with a sound understanding of all that the data imply, the next step requires some creative thought. It is unlikely that any single clear-cut plan can be derived that will accomplish the firm's objectives under the constraints imposed by the market. Given their choice, most contractors—like everyone else—would prefer a large volume of work at a high markup, resulting in an exhorbitant return on investment. Compromises may be required ranging from a small volume at a relatively high markup to a large volume obtainable only with low markups. Some of the different approaches that might be worth experimenting with in this stage include the following:

- Varying the markup according to the size of the job.
- Varying the markup according to the number of competitors on the job.

- Varying the markup according to the characteristics of specific competitors.
- Varying the markup according to the work loads in the industry, guessing how "hungry" the competitors are.

At this point, anything that appears at all reasonable should be proposed.

6.   *Test the Alternatives.*   The alternative plans can be tested as they are developed. Each of the alternatives proposed in the previous step should first be tested against the data collected earlier. Admittedly, testing plans by using the data employed in developing the plans in the first place leaves something to be desired. Still, the *relative* effectiveness of the various plans can be demonstrated in this way. Some of the relationships that can be checked by applying the historical data to various proposed plans include the following:

- The effect of markups on total sales volume. How much volume would have resulted had a 10 percent (or 20 or 30 percent) markup been applied on every job?
- The effect of markups on gross profits. How much profit would have resulted had every job been bid at various markups?
- The effect of markups on net profit margins and the overall profitability of operations.
- The predictability of job outcomes. How do the chances of being the low bidder vary with the markup, job size, and number of competitors?

The efficiency of the various plans can be measure quite simply. The amount of profit that *would have been made* using each plan can easily be compared with the total amount of profit that *could have been made* by taking each job at the lowest competitor's price. *A good plan should enable the contractor to make at least 40 percent as much as he could have made if he had known every competitor's bid in advance.* Many plans will do better. But the only way to find out is to test a large number of plans.

7.   *Select the Best Plan.*   Selection of the best plan must wait until all of the alternatives have been tested against the available data. Having found in the testing stage that many—probably most—of the proposed plans were not spectacularly successful, the alternative plans will eventually be narrowed down to just a few "best" alternatives. From this point, the decision regarding which plan to go with may be extremely difficult. But on the other hand, at this point the only remaining choices are among the relatively good ones, so the worst that can happen is still a good—though not necessarily the best—choice. An

acceptable plan, then, is almost a certainty if everything has been done right so far.

Selecting the best plan requires identifying the one that comes closest to accomplishing the firm's desired objectives. There may be some unpleasant revelations here, though, where *none* of the plans look very desirable, indicating either a poor choice of objectives or the wrong line or business. If the best obtainable results are not good enough, then the contractor must consider very seriously whether he should be in some other type of business. To persist in this case is somewhat like the compulsive gambler who keeps losing in what he knows to be a dishonest card game, just because "it is the only game in town." In this situation, the contractor is offered "Hobson's choice." But in 99 cases out of 100, there is *some* way to do business that will at least come close to achieving the desired results, providing the desires are established objectively—with an eye on the market and competition, as well as on the pocketbook.

8. *Put the Plan into Practice.* As soon as the best plan has been selected, it should be ready to start making money for the firm. Putting it off any longer will be unnecessarily expensive. Before actually placing the strategic plan in operation, though, it should be carefully reviewed to make sure that it is compatible with the established objectives. Since at this point the plan has presumably been checked only with historical data, it will probably be wise to check it on each new job for a while, thereby comparing its results with present policies. If it outperforms anything else available, then it can be officially adopted and used.

9. *Evaluate the Results.* Evaluation of the plan will continue as long as the plan remains in use. The plan's effectiveness can be measured only in terms of how successfully it accomplishes the firm's objectives. This can be measured quite easily, providing the objectives and goals have been adequately defined. Neither objectives nor plans can be considered static or unchangeable. Whenever conditions change—and they *are* changing, continually, in the construction industry—objectives and plans should be carefully reviewed. If the plan is failing to accomplish what it is supposed to, then either the plan or the objectives must be at fault. One or the other, or both, should therefore be changed.

Often, gradual changes will take place in competitors' pricing policies or in the general level of prices in the industry. This may be brought about by an unusually heavy work load increasing the demand for contractors' services, thus limiting their availability and pushing prices upward. Or, a slack period may increase the intensity of competition, with everyone competing on every job and prices approaching the direct out-of-pocket costs of performing the work. In either case the constant state of awareness brought about by the continuing review and evaluation of competitive effectiveness will prove invaluable to the contractor. Minor adjustments can be made almost immediately, and the short reaction time cannot help but result in bigger profits than would otherwise be possible.

Example of a Strategic Plan

The following example is intended only to give the contractor a general idea of what might be expected from his strategic planning. The step-by-step details of how such a plan actually evolves is covered in subsequent chapters.

This example involves a small contractor having annual sales of $300,000 and a net worth of $100,000. His fixed charges (or overhead costs) run $20,000 yearly regardless of volume.

An analysis of what the market is willing to pay and what the competition is willing to work for indicates that a reasonable profit objective for the firm is a 25 percent pretax return on net worth, representing a net profit before taxes of $25,000.

To realize a $25,000 net profit obviously requires a gross operating profit of $45,000 in this case—enough to cover the $20,000 overhead and still leave the $25,000 profit.

Further assuming that total sales will remain at about the same $300,000 level, the contractor can easily calculate the average markup required to achieve the necessary profit as follows:

$$\text{sales volume} = (\text{overhead} + \text{profit})\left(\frac{100 + \text{markup}}{\text{markup}}\right)$$

$$300,000 = (20,000 + 25,000)\left(\frac{100 + \text{markup}}{\text{markup}}\right)$$

$$\left(\frac{100 + \text{markup}}{\text{markup}}\right) = \frac{300,000}{45,000} = 6.67$$

$$6.67\,(\text{markup}) = 100 + \text{markup}$$

$$5.67\,(\text{markup}) = 100$$

$$\text{markup} = \frac{100}{5.67} = 17.6 \text{ percent}$$

On the anticipated $300,000 volume, obtained at an average markup of 17.6 percent on direct costs, the expected year's operations could be summarized like this:

$$(\text{direct job cost}) + (\text{direct job cost})\,(\text{markup}) = \text{total sales volume}$$

$$(\text{direct job cost})\,(1 + \text{markup}) = \text{total sales volume}$$

$$\text{direct job cost} = \frac{\text{total sales volume}}{1 + \text{markup}}$$

$$\text{direct job cost} = \frac{300{,}000}{1.176} = \$255{,}000$$

A summarized operating statement, then, would look like this:

Total sales volume	$300,000
Less direct job costs	255,000
Gross operating profit	45,000
Less overhead expenses	20,000
Net profit before taxes	$25,000

The $255,000 in direct job costs was found by taking the $300,000 sales volume and dividing by 1.176.

Having established this objective, then, the problem is to devise a plan that will accomplish it. Sparing the details (which will be covered later), the following two-part strategic plan might be developed. The first part of the plan deals with job selection, identifying *which* jobs are to be bid; the second part sets the pricing policy, telling *how much* to bid on each job.

Which Jobs to Bid

Considering both the chances of getting jobs of varying sizes and competitive characteristics and the amount of profit to be made by getting them, the most advantageous conditions for bidding can be determined. Figure 3 shows how the priority chart might look for a typical operation.

Figure 3 breaks out seven priority classes according to the job size and the number of competitors. These classes are numbered consecutively from 1 (the best) to 7 (the worst). All the jobs listed in a single category are roughly equivalent in terms of their attractiveness from a profit standpoint. This implies, for example, that the profit to be gained in the long run from bidding $30,000 jobs against a single competitor is roughly comparable to what would be expected from bidding jobs in the over-$50,000 class against four or five

Job Size	Number of Competitors				
	1	2	3	4	5
$5000 and under	VI	VI	VII	VII	VII
$5000 to $20,000	IV	V	V	V	V
$20,000 to $50,000	III	IV	IV	IV	IV
$50,000 and over	I	II	II	III	III

FIGURE 3. Example of priority system: priority class.

competitors. In developing such a priority chart, nothing can be taken for granted, although it is probably safe to assume that bidding a large job involving little competition is preferable to bidding a small job against many competitors. Aside from these two extremes, though, the priority groups require some careful analysis.

These priority groups simply indicate that, given a choice of several jobs, those having the highest priority rankings should generally be bid first. However, as shown in the following section, limiting the firm's bidding to a single category of job—even the top-priority group—is not desirable and could easily prevent the firm from realizing its objectives.

How Much to Bid

Figure 4 indicates how much of a markup should be included in the bid for each job, again depending on the job size and the number of competitors invovled. The breakdown is the same as on the job priority chart. As indicated in Figure 4, the markup would be highest on small jobs bid against few competitors, and lowest on large jobs bid against many competitors. The number of competitors has a far greater effect on the markup than does the job size.

The optimum markups shown in Figure 4 can be determined only after a detailed analysis of the contractor's past jobs. They assume no knowledge on the part of the contractor other than the job size (which he will know) and the number of competitors (which he may know, or may have to estimate based on his experience and judgment).

In the example, to achieve his target of an average 17.6 percent markup, the contractor obviously expects to obtain jobs of several different types. If his estimated $300,000 volume were made up of just three $100,000 jobs, each bid against a large number of competitors, at the indicated 9 percent markup he would of course fail to meet his profit objective for the year's operations. He would, in fact, barely recover his overhead expenses. His financial summary would look like this:

Total sales volume	$300,000
Less direct job costs	275,000
Gross operating profit	25,000
Less overhead expenses	20,000
Net profit before taxes	$5,000

This $5000 net profit represents a miserable 5 percent return on net worth and a paltry pretax profit margin of only 1.7 percent on total sales.

If the contractor in question were able to obtain the same $300,000 volume spread equally over the different priority groups, though, he might end up like this:

Priority Group	Average Markup (%)
2	14
3	16
4	17
5	16
6 and 7	25
Average	17.6

To get the markup required by his established objectives, then, requires obtaining a number of jobs in the lower priority groups which carry relatively high markups because of their size or competitive characteristics.

By setting his objectives on a different basis, this contractor could probably improve his performance. Budgeting his operations for the same 25 percent return on net worth but on a $400,000 sales volume with an average markup of

	Number of Competitors				
Job Size	1	2	3	4	5
$5000 and under	45	28	23	17	14
$5000 to $20,000	34	23	17	14	11
$20,000 to $50,000	28	18	14	11	10
$50,000 and over	23	16	12	10	9

FIGURE 4.   Example of pricing system: optimum markup.

12.7 percent would have permitted him some added flexibility in job selection without sacrificing any profit opportunities. His markup schedule would be the same, but in this case he would have "turned over" his capital at a faster rate. The key to improving performance would lie simply in bidding jobs with higher profit potential.

Regardless of how he sets his objectives, priorities, and pricing schedules, he will always be looking for the "plums"—the large jobs attracting little competition. He can always revise his objectives upward if he is successful in finding enough of these opportunities to profitably increase his volume.

Summary

Strategic planning refers to the procedures for developing and achieving specified business objectives. For the contractor, strategic planning is concerned primarily with identifying *which* jobs to bid and *how much* to bid on them. A competitive strategy offers the contractor many benefits in attaining the maximum profits possible under existing competitive conditions. The overall strategic plan consists of four major elements: (1) objectives, (2) the business plan, (3) goals, and (4) monitoring. These elements evolve through the strategic planning process, which includes nine distinct steps: (1) identify the problem; (2) define the objectives; (3) collect the data; (4) interpret the data; (5) devise alternative plans; (6) test the alternatives; (7) select the best plan; (8) put the plan into practice; and (9) evaluate the results. A logical and effective strategic plan can be developed by any contractor, regardless of his scale of operations, provided he follows these steps. The end result of the planning process will be (1) a priority system, identifying the relative desirability of different types of jobs, and (2) a pricing system, identifying the markups that will result in the greatest possible profits in a given situation. Experience has shown that a good strategic plan will enable the contractor to make at least 40 percent as much profit as he could have made if he had known every competitor's bid in advance.

4

The Contract Construction Industry

*When a business firm attempts to mold its whole
policy to meet the prices of its competitors that business is entering a labyrinth, the center of which is the
chamber of despair. Highest quality never can be
given nor obtained at the lowest prices. If a price must
be sacrificed, quality must be sacrificed. If quality is
sacrificed society is not truly served.*

H. T. GARVEY

The construction industry has many unique characteristics which distinguish it
from other industries. The industry's recent era of "profitless prosperity" offers
dramatic evidence of some of these unique features.

The construction industry as a whole has been plagued by low profits and
intense competition, by a high rate of contractor turnover and failure, by
restrictive legislation and binding labor agreements, by antiquated building
codes, and by a general lack of research.

Still, the industry is, without doubt, composed of the most rugged group of
individualists in American industry. This is probably because mere survival in
the construction industry requires a high degree of individuality and rugged-
ness, and those who do not possess the necessary attributes are unable to stay in
business for any length of time.

Composition of the Industry

By definition, a contractor is one who acquires an obligation to perform work
or to supply materials or a finished product on a large scale at a specified price.

33

This definition covers many operations outside the construction industry as well as those within.

The construction industry alone includes at least several dozen different types of contractors, each specializing in a particular phase of construction activity. Altogether, there are roughly 500,000 firms engaged in some form of contract construction in the United States. The construction industry can be broken down into the following major categories:

1. General building construction.
2. Building trades.
3. Engineered construction.
4. Other specialties.

General Building Construction

General building contractors are firms engaged in the construction of all types of residential, commercial, and industrial buildings; they include both general builders and speculative builders. Speculative construction usually refers only to the residential field, where the developer-contractor builds houses before he has a buyer for them. In the speculative operation, the firm must decide on the scope of its operation, acquire land, develop plans, construct the houses, and hope to sell them when they are finished. This type of operation requires a relatively large amount of capital for the dollar volume of sales generated and carries with it a high degree of risk; it is therefore limited to relatively small projects. Most general builders outside the residential field bid or negotiate only on jobs for which plans and specifications have already been prepared by architects or engineers, and on which they have been invited to submit proposals or bids.

Building Trades

The building trades are the domain of subcontracting, a highly specialized and fiercely competitive field. The subcontractor usually works under contract with the prime contractor on a job. The major categories of building subcontractors embrace heating, plumbing, air conditioning, painting, papering, electrical work, masonry, stone work, plastering and lathing, terrazzo and tile, carpentering, flooring, roofing, sheet metal work, concreting, general building maintenance, structural steel erection, ornamental iron and steel work, glass and glazing, excavation, foundation, wrecking, moving, machinery and equipment installation, and a host of other special skills.

Engineered Construction

The engineered construction category covers, generally, works that are planned and designed by engineers rather than architects. Typical projects falling within this classification are further divided into highway and heavy groups, and even more specifically into bridges, grading, municipal and utilities (water mains, sewers, etc.), paving, surfacing, dams, and other specialties. A particularly important category in the engineered construction field is electric power plant construction.

Other Contractors

There are numerous other classifications of contractors which are of particular importance in certain areas. These include marine contractors, oil-well and drilling contractors, pipeliners, industrial and chemical process plant constructors, and many others.

Despite the abundance of construction specialties, all contractors have many problems in common due to the close similarity in their methods of operation.

Comparison with Other Industries

A certain amount of risk and uncertainty is a normal part of nearly every business enterprise. One important way in which the construction industry differs from others is in the degree of uncertainty. The manufacture's labor force is fairly constant, his products are uniform in their specifications, his costs of production can be identified and controlled, the markets for his products can be tested, and pricing policies can be established. The contractor, however, must employ temporary labor, have his work schedules halted or altered by adverse terrain and weather conditions, continually perform different types of jobs according to different specifications, experience widely (and wildly) fluctuating costs, be subject to a complete lack of price stability, and face an unknown future market.

In some respects, then, it is not surprising that the contracting business suffers in comparison with other industries, especially when the comparison is in regard to profits. While the specific figures change from year to year, a comparison among some of the major industrial groups can be revealing. Table 6 shows how recent profits in the construction industry compare with those in several other industries.

TABLE 6.   Relative Profitability of Selected Major Industries

	Profit as a Percentage of Sales	Ratio, Sales to Assets	Profit as a Percentage of Assets
Contract construction			
General building contractors	1.1	3.3	3.6
Heavy construction	2.0	1.4	2.8
Electrical contracting	2.4	1.7	4.1
Plumbing, heating, air conditioning	2.1	2.8	5.9
All contract construction	1.9	2.0	3.8
Manufacturing	5.2	1.2	6.2
Agriculture	5.6	1.2	6.7
Mining	14.4	0.8	11.5
Public utilities	5.9	0.5	3.0
Wholesale trade	2.5	2.7	6.8
Retail trade	3.3	2.7	8.9
Business services	17.5	1.2	21.0

Types of Contracts

The most common types of contracts under which contractors work are the lump-sum contract, the unit-price contract, and the cost-plus contract, plus different combinations of these three.

The Lump-Sum Contract

The lump-sum contract is by far the most common type of contract for construction work and may be awarded either through competitive bidding or by negotiation with the owner or contracting agency, with most large jobs coming from competitive bidding. In the lump-sum contract, the contractor agrees to perform the designated work as specified for a fixed price. If the cost exceeds the price, there is no recourse for the contractor, who must pay the difference himself. If he can complete the work at less cost than was estimated, the difference is in his favor.

The Unit-Price Contract

Unit-price contracts are, as their name implies, based on a specified price per unit of work performed. They are particularly applicable to jobs where the units of output can be easily measured, but cannot be estimated at the time the job is let. Rock excavation, for example, might be appropriately included in a unit-price contract when the amount of rock to be encountered cannot be determined in advance. The same is true of oil-well drilling and some phases of pipeline construction where a price per foot might apply better than a fixed total price.

The Cost-Plus Contract

Cost-plus contracts are frequently used when the work cannot be defined in advance, for such a contract may save time and increase flexibility in these cases. The "plus" portion of the cost-plus contract can be expressed either as a percentage of total direct cost or as a fixed fee to be added to the direct job cost. The cost-plus contract works well when the owner and contractor have complete faith in each other, but financial difficulties and ill feelings sometimes are caused by this system. The owner from the very beginning is uncertain about his costs for the job, and there is little incentive for the contractor to keep his costs down.

Problems of the Competitive Bidding System

Most major construction work is obtained through competitive bidding; this practice has been generally criticized by many contractors as the basic problem of the construction industry.

If the contractor bids low enough to get a job, he cannot make a fair profit. If, on the other hand, he bids high enough to make a fair profit, he may be unable to get a job. These unpleasant alternatives place the contractor in an extremely awkward position.

While the awarding of contracts for construction work on the basis of competitive bids offers certain advantages to both owners and contractors, many of the construction industry's problems can be attributed directly to the practice of making price the sole criterion. The lowest price for a job frequently means, naturally enough, that the job will be carried out under the lowest standards possible in view of the amount and strictness of supervision and inspection.

Under the competitive bidding system, the contractor is forced to make a "short sale" of his resources. In effect, he is selling a finished commondity—a building, for example—which he does not have and which does not even exist at the time the sale is made. The contractor is gambling that he will, within a prescribed time, be able to furnish the end-product at the price orginally set. To do this, he must anticipate a number of uncertain, unknown, and often uncontrollable circumstances; however, he dare not reflect these contingencies in his bid price or the job will almost certainly be awarded to a less cautious competitor.

Under these conditions there is little wonder that profit margins are low in construction contracting.

And the element of uncertainty is but a small part of the total problem. Probably the greatest single factor affecting profits in the construction industry is the degree and type of competition brought about by the contractors themselves. There are two kinds of contractors who, although personally convinced of the soundness of their business philosophies, probably do more harm to the industry than any other. These are (1) the price-cutter and (2) the contractor who bids every job.

The Price-Cutter

The impact of the low-markup, high-volume business philosophy on the construction industry cannot be measured accurately, but millions of dollars in profits are undoubtedly sacrificed each year because of this type of operation. A job taken at a price closely approaching the direct job cost benefits no one. The contractor who gets the job makes nothing at his minimal price, and those who bid the job without getting it merely incur the additional costs of preparing their unsuccessful bids. The net effect on the industry as a whole is worse than if the jobs were never offered in the first place, for it results only in additional costs with no compensating profits.

The "Bidding Fool"

Contractors who feel obligated to bid every job that comes along also contribute much to the profit squeeze by increasing the intensity of competition on each job they bid, thereby lowering the price at which the job is finally let. It is not only possible, but often happens, that as total construction volume increases and the number of contractors decreases, competition becomes even more intense than before. Only the average number of bids per job need increase to bring about this situation, the result of each contractor increasing the number of bids he submits. The intensity of competition encountered on a job depends solely on the number of bids on that one job, not on the total number of contractors in business. And the chances of a low-side estimating error on a job are, of course, substantially increased as the number of bidders increases.

Contractors, then, have brought on themselves many of the problems blamed on the competitive bidding system. There has always been, and will probably always be, sufficient work to allow everyone a fair profit, providing everyone refuses to settle for anything less than a fair profit on every job. This, rather than the "system" itself, is the main problem, for there is always someone who is willing to settle for a little less than an adequate return.

Contractor Turnover

A very high mortality rate prevails in the construction industry. Each year, more than 1000 firms in existence at the beginning of a year fail to make it to year-end. Meanwhile, however, they are replaced by even more new firms just entering the field.

Data for recent years, for example, show that there are over one million firms engaged in contract construction in the United States. During each year, some 27,000 new firms come into existence, while 1700 existing firms fall into bankruptcy. Overall, the contract construction industry accounts for between 8 and 10 percent of the nation's gross national product, and for just over 8 percent of the total number of business firms in the United States, but accounts for more than 17 percent of all business failures.

One reason for the high rate of failure in the construction industry is the relative ease of entry into the business. A journeyman tradesman with a pickup truck and enough money to support his family for several months can often obtain sufficient credit from suppliers and a bid bond from a bonding company so that he can bid a number of sizable jobs, even though his financial resources and managerial abilities are perhaps inadequate to handle the jobs he bids. Still, with his low overhead and even lower markups, he makes difficult competition for older, well-established firms, and he may be able to obtain enough jobs to continue in business. Many contractors in this category fail to acknowledge that they have *any* overhead expense; all income over and above the direct out-of-pocket cost of labor and materials is considered profit. And sometimes, what he considers "profit" is simply journeyman wages (or less) for his own time spent on the job.

Once he is able to get a few jobs, though, the marginal operator can stay in business for a long time without ever making a legitimate profit. Suppliers and bonding companies are understandably reluctant for him to fail, thus defaulting on his outstanding acounts; hence he can continue receiving easy credit. As long as he can keep work coming in (at any price), he can keep paying on his previous month's debts with his current month's receipts. Until, that is, the present month's receipts fail to materialize and he becomes another statistical entry in Dun & Bradstreet's failure record.

Effect of Contractor Failures

Whenever a contractor fails, everyone in the industry loses or has already lost. For example, if a contractor were to fail leaving total liabilities of, say, $50,000, this $50,000 liability was probably accrued over a lengthy period of time covering many jobs. This liability might have resulted from a $10,000 deficit suffered on each of five different jobs; if each of the five jobs represented a $200,000 contract, a million-dollar volume was sacrificed. Actually, well over a million dollars' worth of work might well be represented by liabilities of this amount, or even of a much smaller amount. And when a situation like this occurs, the volume of work that was involved might just as well never have been on the market as far as the construction industry as a whole were concerned, for the industry as a whole gained nothing by it—just a large, profitless volume.

When this liability is, in turn, multiplied by the several thousand contractor failures that occur annually, the effect on the construction industry is very significant—amounting to as much as several billion dollars of potentially profitable work each year from which the construction industry as a whole is unable to benefit.

It is apparent, then, that the cost of contractor failures is paid for by the entire industry.

Causes of Contractor Failure

The causes of business failure are more easily identified than corrected. Nevertheless, there is a great deal to be learned from analyzing the causes of failure. By avoiding these pitfalls, success will still not be assured but it will at least be within reach.

Dun & Bradstreet has compiled much information on the causes of failure in different industries. The construction industry has contributed more than its share to these statistics; with slightly more than 5 percent of the nation's total nonagricultural employment, the construction industry generally accounts for over 17 percent of all business failures.

Dun & Bradstreet lists the major causes of contractor failure in the following order, with the first three items alone accounting for more than 80 percent of the industry's failures.

1. Incompetence.
2. Lack of managerial experience.
3. Unbalanced experience.

4. Lack of experience in the line.
5. Neglect.
6. Fraud.
7. Disaster.

Almost all these causes of contractor failure can be attributed to various short-comings on the part of management. The first four underlying causes of failure, for example, reveal themselves in other, more obvious ways—inadequate sales, excessively high operating expenses, receivables difficulties, inventory problems, excessive capital tied up in fixed assets, competitive weaknesses, and other problems.

Incompetence

Incompetence covers a multitude of evils. The term is defined in the dictionary as "wanting in adequate strength, capacity, qualifications, or the like; one incapable of managing his affairs because mentally deficient or undeveloped." Incompetence may refer to the contractor's lack of either managerial or technical capabilities, but experience has shown that managerial inability is far more common and far more disastrous from a financial standpoint than are technical deficiencies.

Lack of Managerial Experience

Many failures in the construction industry occur in firms managed by men who have worked their way up through the building trades. While such a back-ground does provide excellent training for many management responsibilities, the business shrewdness required of construction management is not necessarily acquired during a trade apprenticeship, and there is little opportunity to gain the necessary business skills through practice later when real money is at stake. Trial-and-error business experience is quite expensive, especially if there are many errors.

Unbalanced Experience

Unbalanced experience refers to experience which is not well rounded in sales, finance, accounting, purchasing, production, estimating, office administration, and all the many other skills required of construction management. There are relativey few individuals who possess all the know-how required for all phases of the contracting business, and prudent delegation of responsibility for many of these functions is therefore essential.

Lack of Experience in the Line

People outside the construction industry sometimes think of contracting as an easy way to make money. Many contractors share this view, when applied to types of construction work with which they are not familiar; invariably, they feel that better profit opportunities exist in fields other than their own specialties. Consequently, when contractors who specialize in certain types of work begin to feel the effects of competition in their own line, they often try to branch out into other fields of construction. This increases the competition for firms in these other fields, increases the chances of errors, and frequently results in complete financial disaster for the inexperienced.

Neglect

Neglect is the result of human weaknesses. According to the opinions of informed creditors and information in Dun & Bradstreet's credit reports, the major factors contributing to neglect include bad habits, poor health, and marital difficulties, with poor health the most common.

Fraud

Outright and intentional fraud, as evidenced by false financial statements or irregular disposition of assets, is less of a problem in the construction industry than in most other industries.

Disaster

Contractors place their capital at the mercies of the gods on every new contract. Even so, disasters, including fires, floods, burglaries, employees' fraud, and strikes, account for very few contractor failures. Perhaps this is because, unlike managerial deficiencies, disasters are general nonrecurring.

Other Industry Problems

The construction industry is not without its other problems, in addition to those brought about directly by the competitive bidding system. Many of these problems are simply the result of the industry's unique characteristics. Because of the relative ease of entry into the contracting business, there is an uncommonly large number of marginal operators; this in itself results in an uncommonly large number of problems. Just a few of the construction industry's major problems, outside the individual problems of specific contractors, include bid peddling, retainage, and a general lack of research.

Bid Peddling

Most contractors are aware that even having the lowest bid on a job will not necessarily assure them of getting the job. Bid shopping (or peddling) refers to the practice of taking competitive bids, then offering the job to a preselected contractor or subcontractor at a price just below the low bid. While bid shopping is admittedly a serious problem, its victims are usually as guilty as its perpetrators, the procedure requiring the cooperation of at least two parties. Different methods have been employed to combat bid peddling. The most effective way is simply to not bid those jobs on which shopping is expected. Another common practice is to wait until the last possible moment before the letting to submit bids, the thought being that there will not be enough time for shopping; this approach, however, creates other problems. A third alternative which has been employed successfully by many groups is the bid depository, in which the competing contractors deposit records of their bids with a single, neutral party.

Retainage

Retainage refers to some percentage of the amount of money due the contractor for work that he has already performed; the money is held back or retained by the owner to insure satisfactory completion of the project. The usual retainage on jobs, accompanied by a frequent slowness in paying contractors and subcontractors, has imposed considerable financial hardships on many contractors and their creditors. Frequently, the amount of retainage is greater than the total markup on the job, and firms with limited capital are therefore unable to meet their obligations when due.

Lack of Research

Research is one of the major areas in which the construction industry is deficient. Construction has traditionally been a craft-based industry rather than a research-oriented one. This lack of research activity is due primarily to the structure of the industry, with its large number of relatively small and highly competitive firms providing no single, broad, unified base for supporting research. A single firm cannot afford research unless it will pay off immediately, for its cost can seldom be written off in the bid. Consequently, most research on construction methods, materials, and procedures has been carried on by construction equipment manufacturers and materials producers.

Under the principles of free competition prevailing in the construction industry, the opportunity to succeed is also an opportunity to fail. It is indeed unfortunate that so many contractors avail themselves of the latter opportunity.

Summary

Contractor failures and low profit margins in the construction industry indicate a strong need for improved management effectiveness. Many problems exist in the industry, some of which are caused by the contractors themselves, and others of which are brought about by the industry's unique characteristics. The construction industry is composed of many different types of contractors, each with a specialty of his own. Even so, all contractors share similar problems caused by the competitive bidding system and by the irrational behavior of misinformed competitors. Contractor failures, almost invariably the result of management weaknesses, cost the industry as a whole millions of dollars in profits each year. Only through a better understanding of sound management principles on the part of all contractors can these problems be completely overcome.

5

The Requirements for Success

*On the clarity of your ideas depends the scope of
your success in any endeavor.*

JAMES ROBERTSON

Success is sometimes said to be a journey, not a destination; a goal, not an achievement. Success is never final but must continually be pursued.

The requirements for successful contracting are, in some respects, similar to the requirements for successful golf. In order to accomplish his objective—getting the ball into the hole, in this case—the golfer must first know where to start; next, he must know where he is going; and, finally, he must know the best way of getting from where he is to where he wants to go. This knowledge is essential prior to attempting to carry out his plan. In working toward his objective, the golfer must perform many separate functions concurrently and correctly. Should only one serious error be made (such as failing to "follow through," for example), the player is likely to be penalized by ending up in some undesired location.

In most human endeavors, whether sports or business, there are more wrong ways to go than there are right, and the winners are usually those who make the fewest mistakes—in other words, the winners are those who do not beat themselves. This is most certainly true in the contracting business.

Simply stated, the primary questions to be answered in plotting the strategy for a firm's success are as follows:

1. Where is it now? The present business must be analyzed as it currently stands, in order to make it as effective as possible under present conditions.
2. Where is it going in the future? The firm's potential must be assessed realistically, and its future objectives formulated.

45

3.   How will it get there? Plans and procedures must be developed which will enable the firm to progress from its present status to its desired future position as effectively as possible under the expected conditions.

This chapter deals primarily with the first point: how to determine the strengths and weaknesses of the present organization, some of the general requirements for making it more effective, and the responsibilities of management in employing the firm's resources and capabilities to maximum advantage.

The Functions of Management

Success in business is the evidence of good business management. The success of the contracting firm depends primarily on the ability of its management to carry out its responsibilities efficiently and effectively. The construction executive's major management functions cover eight broad areas encompassing a wide range of skills and requiring a high degree of flexibility. These management functions include the following:

1. General office administration.
2. Personnel administration and labor relations.
3. Design and engineering.
4. Operations.
5. Finance and accounting.
6. Legal.
7. Business planning.
8. Sales promotion and management.

General Office Administration

Many routine chores are associated with running any type of business. But if the bulk of these clerical-type duties are not delegated to clerical personnel, the construction executive may become so bogged down in minor details that he will be unable to keep up with other responsibilities that cannot or should not be delegated. A tremendous amount of paperwork is involved in taking care of routine correspondence, records, and reports; however, the manager's job is to *use* these reports to control and improve the business, and their preparation can generally be handled by less expensive personnel.

Personnel Administration and Labor Relations

The contractor must be particularly adept in his dealings with subordinates, for effective use of manpower is necessary before maximum benefits can be obtained from the other three resources—machines, materials, and money. Personnel duties include finding and hiring good people, plus training and holding on to them. Contract negotiations and the handling of labor grievances and disputes are also critical phases of the contractor's personnel administration program.

Design and Engineering

Although most contractors may never be called on for actual design of the projects on which they work, a sound understanding of the project's technical aspects is very important. Most engineers and architects would probably be better in their own professions had they some field experience in construction. Similarly, most contractors could improve their work and avoid many problems with a better knowledge of engineering and design principles.

Operations

Management of construction operations is, to most contractors, the most fascinating part of the business, sometimes to the point of neglecting other equally necessary phases. Management's responsibilities in the construction operation involve allocating and scheduling the company's financial and physical resources. Operations, in its broadest sense, includes all phases of the construction project, from the initial selection of jobs to be bid, through estimating, bidding, determining the overall approach to be taken in the work, scheduling manpower and equipment, procuring materials, letting subcontracts, coordinating all parts of the project, controlling costs, enforcing adequate safety precautions, and handling many other job-related details.

Finance and Accounting

The need for sound financial management is always present. Management's primary responsibilities with regard to financial matters include both obtaining and using capital effectively to enable the company to operate with maximum profitability. To accomplish this, financial transactions must be recorded, summarized, and interpreted. While the compilation of financial data can be delegated to others, management is still responsible for the interpretation and use of these data; effective use of financial information requires that management possess a sound fundamental knowledge of the principles of financial management.

Legal

The adage, "one who acts as his own attorney has a fool for a client," has a great deal of truth. Contracts and specifications, being legal documents, should be examined by a competent lawyer, for in many cases the contractor could be financially ruined if he were required by the architect or engineer to comply strictly with the specifications. And big losses almost always result when the contractor is forced to employ legal measures to collect on his accounts. While having a lawyer is a poor substitute for staying out of trouble in the first place, it can help to avoid being completely overwhelmed by legal complications.

Business Planning

It is far better to have a poor plan than to have no plan at all. But it is better yet to have a plan that has been well thought out with specific goals in mind. Good business planning indicates what action needs to be taken now to help achieve future objectives, helps management to coordinate the different phases of the business, and provides an effective basis for controlling operations. Business planning includes setting objectives, establishing policies for achieving the objectives, developing an overall strategy for operating the business, defining procedures for delegating authority, and formulating short- and long-range plans.

Sales Promotion and Management

Insufficient sales volume is probably blamed for more contractor failures than any other single factor, although it is generally the result of poor sales planning and is perhaps further attributable to a general lack of knowledge concerning the amount of work actually needed to recover overhead costs. That most sales come from offering to do the maximum job at the minimum price results in generally low markups, which in turn results in high volume requirements. It is management's responsibility to gear sales ojectives to the existing market structure and competitive conditions.

The Principles of Successful Contracting

Most successful contractors attribute their good fortune to an abundance of shrewd business skills. Those who have fared less well, on the other hand, are equally certain that their state has resulted from bad breaks and hard luck.

Successful contractors are frequently subjected to lengthy explanations from their less successful counterparts who are continually "growling and grumbling and taking it easy."

The contractor who succeeds is the one who knows his capabilities, establishes his objective, and works his way step by step toward it. The following 13 general principles have been suggested by a number of successful contractors as being particularly important in running a profitable contracting business:

1. Be selective in choosing jobs to bid.
2. Before bidding a job, analyze it carefully and objectively.
3. Exercise particular care in estimating.
4. Include a realistic markup in all bids.
5. Establish effective cost control and cost reporting procedures.
6. Maintain adequate cash reserves and working capital at all times.
7. Employ the best supervisory personnel available.
8. Control overhead costs carefully.
9. Develop a strong technical competence and a knowledge of the market to be served.
10. Maintain good public relations.
11. Be alert for ways to improve the operation.
12. Use subcontracts for specialized work.
13. Try to obtain at least part of the work through negotiation.

Job Selection

The contractor has an opportunity to bid on hundreds of jobs each year, many more than he has either the time to estimate or the capacity to perform. From the vast number of available jobs he must choose the relative few that he will bid. And since bidding a job necessarily entails considerable expense in both time and money, it is imperative that jobs be selected that offer at least a fair chance of earning a profit. For example, several jobs are apt to be available at the same time, but limitations in time or personnel may mean that only one can be bid; the contractor must therefore be careful to choose the one that offers the best profit potential. Indiscriminate bidding is not only costly, it is disastrous— both for the guilty contractor and for the industry as a whole.

Job Analysis

The best of architects' and engineers' plans and specifications are sometimes deficient or unclear, and unexpected conditions, whether encountered in the job itself or in the interpretation of specifications and contract documents, can easily wipe out any chance for a profit. It is far better to anticipate the unexpected before it happens than to fight it afterward. Every job carries associated risks; when the risks outweigh the potential profits on the job, however, the job had best be left alone.

Estimating

Success often depends on the contractor's ability to make accurate estimates of total job costs; even the simplest jobs entail enough risk to warrant extreme care in estimating. Every contractor knows the effect of errors in estimating. These errors are invariably on the low side, since contractors erring on the high side seldom get the job, and the true costs are never revealed. Low cost estimates simply mean that costs must be recovered through the markup, thereby resulting in profitless volume.

Bidding

The most careful job selection, analysis, and estimating are to little avail unless the contractor is also able to get the job with an adequate markup. This is always difficult and is frequently impossible. The best bid for a particular job is the bid which, in the long run, can be expected to result in the best possible combination of (1) the profit to be made by getting the job at a specified price, and (2) the chances of getting the job at that price. The best bid on a given job is not necessarily the low bid; in fact, if a contractor were to take every job he bid at the low bidders' prices, he would probably soon be bankrupt. The only way to make a profit is to bid every job at a profit; trying to get volume just for the sake of volume is a firm invitation to financial disaster.

Cost Control

An effective means of recording and reporting costs is necessary to keep costs in line with the original estimates, and to relate the work completed—the part of the operation that brings in money—to the costs incurred. Detailed cost reporting and control systems also help focus attention on opportunities for savings in future jobs, and are useful in developing and updating estimating and performance data for future use.

Adequate Financing

A lack of adequate capital causes many businesses to fail, and prevents many others from growing. One of the main causes of financial difficulty is when operations have been expanded to a point where no capital reserve is available to cover obligations when they fall due. A firm can easily expand itself into bankruptcy by being unable to pay its bills until it collects on all of its own outstanding accounts. Adequate capital helps to avoid this problem and also helps the business in other important ways:

1. By minimizing the extent to which credit need be used, thereby picking up additional discounts.
2. By providing capital for the purchase of cost-cutting equipment.
3. By furnishing a "cushion" for carrying the firm through slack periods.
4. By reducing the pressure or temptation to get jobs at any price, just to keep money coming in.
5. By qualifying the firm to bid bigger and better jobs, and enabling it to be more selective in choosing jobs.

Overhead Cost Control

The relationships among overhead costs, direct costs, markups, and sales volumes should be thoroughly understood by every contractor. A high-profit operation can be achieved with many different combinations of these four elements, but overhead costs can be more easily controlled by management and can actually have a greater and more direct effect on profits than can the other three. Each dollar of overhead cost incurred in the office must be offset by many more dollars of income from operations in the field with the multiplier depending on the level of markups. High fixed costs require either a high volume of work at normal markups or higher markups on a lower volume; either is difficult to obtain in competition with contractors who are not thus handicapped.

Knowledge of the Market

To be successful, the contractor must first have a strong technical competence in his chosen specialty and a clear picture of the role of his own operation in the total industry structure. The volume of construction activity varies widely from year to year and even from month to month in every area. In recognizing the overall opportunities for profitable work, the contractor must be alert to trends in both the volume and type of construction activity and be able to relate these trends to his own business. Careful observation and analysis of the actions of competitors is an important ingredient of the firm's market strategy; objectively recognizing the impact of the competitive situation on the existing construction market can result in substantial savings and ultimately in greater profits.

Public Relations

Maintaining good relationships with owners, architects, engineers, public officials, and labor officials, is necessary in assuring smooth operations. The best way to have prospective clients think highly of the contractor's work is, of course, for the contractor to do work that is highly thought of. In an intensely competitive business such as contracting, what others think can often spell the difference between success and failure. A reputation for good work, plus friendly relations with owners and their representatives, will result both in more invitations to bid and in more opportunities for negotiated work.

Improving Operations

New construction equipment, materials, and designs affect construction methods and costs, and operations must be continually adapted to take advantage of the changes. While it is seldom wise to be the first to discard old methods, it is equally unwise to be the last. A competitive advantage can frequently be obtained, and sometimes retained, by being among the first to adapt new techniques to existing problems. The construction press offers a wide variety of useful publications for contractors who are willing to take the time to learn the new concepts and their applications. Those who fail to keep up with new developments in their field may soon find that their field has left them behind, for the contracting business has no place for those who are unwilling to learn.

Subcontracting

Profitable contracting depends to a large extent on the proper use of subcontracting. Few construction firms have or can afford to support all of the capabilities required to handle even relatively small jobs. Subcontracting is a highly specialized field, and subcontractors are usually able to offer considerable savings in both direct and administrative expenses when compared with the alternatives of staffing the firm with the necessary skills on a full-time basis.

Negotiated Work

Efforts should be made to balance the volume of work obtained through competitive bidding with work obtained through negotiated contracts. While most government agencies are practically forced to accept the lowest price offered on a job (although frequently accompanied by a sacrifice in the quality of workmanship), many business and industrial firms are aware of the dangers of the lowest price; they can, and do, award many of their construction contracts on a more selective basis, often through negotiation. This is a type of work which should be sought actively by contractors, by developing the broad experience and effective organization required to work closely with owners, architects, and engineers, and to best meet the needs of prospective clients.

Case Histories of Successful Contractors

One common characteristic of successful contractors stands out above all others: they are *profit-minded!* They would rather make a large profit on a small job than a small profit on a large job; they would rather do a relatively low volume of work at adequate profit margins than take on huge projects at the low markups necessary to obtain these jobs through cutthroat competitive bidding. Generally, the most successful contractors are neither more nor less technically competent than others, but their attitude toward profit separates them from their less successful, volume-oriented competitors. The following four examples describe briefly how contractors in several different fields have achieved an enviable degree of success. These examples are not unusual or spectacular; similar examples can be found in nearly every area and in every specialty. Every contractor, by emphasizing profits rather than volume in his own opera-tions, can become equally successful.

Case 1: General Contractor

This contractor, working in the highway-heavy field, and particularly in its highly competitive river and harbor improvement phases, is a good example of an operation that changed from near failure to exceptional profitability simply by adopting a more discriminating approach to bidding. In its race toward bankruptcy, the firm bid nearly every job that was offered; and since the same jobs were offered to many other contractors, the bidder's lists sometimes contained 20 or more names. Bidding such a large volume of work naturally entailed considerable administrative expense, including a large staff of estima-tors. Management's insistence on "bidding the job, not the competition" simply resulted in getting only a few jobs, mostly when only a few competitors were involved. This combination of high overhead costs and low volume finally made changes necessary. Fortunately, management chose to cut overheads rather than prices; this in turn meant that fewer jobs could be estimated and bid; in choosing the smaller number of jobs out of the many jobs available, a more selective approach was possible. This practice paid almost immediate dividends. By analyzing past data, the contractor identified the characteristics of particularly successful projects—types of jobs, geographic areas, competitive conditions, contracting agencies, and so on—and similar characteristics were subsequently looked for among the currently available jobs. For example, it was found that only one job during the past three years had been won when more than 12 competitors were encountered, and further checking indicated that there had been an error in the estimate on that job. Consequently, jobs that were likely to draw 12 or more competitors were no longer considered; this practice immediately eliminated nearly half of the jobs that might otherwise have been bid—along with half the total bidding expenses, but with no loss of profits! Profits soon improved, resulting from a combination of lower overhead costs and a greater proportion of successful bids. And the initial successes brought about more success; the company was no longer in the position of *having* to bid so many jobs just to cover costs and was financially able to survive slack periods.

Case 2: Building Contractor

This firm is perhaps unique among contractors, not only in its consistently high level of profits, but in its method of securing business: essentially all of its volume, averaging nearly $10,000,000 annually over the past five years, is in negotiated private work, mostly industrial, commercial, and institutional buildings. In more than 20 years of successful operation, the firm has built a reputation for experience, quality, and performance that has largely eliminated the need for competitive bidding. For example, this contractor was recently awarded through negotiation a contract for construction of a $20,000,000 institutional complex simply because the developers were so firmly convinced of the firm's integrity and ability to give them the job they wanted. The key to this contractor's success, then, can be summed up in just a few words: good performance and good public relations. Although the owner's background includes an engineering degree and on-the-job construction experience as estimator, foreman, and superintendent—all important in themselves—his most profitable talents obviously lie in the field of customer relations. His firm sells itself as strong, capable, experienced, and interested in giving the client the most for his money. The customer's confidence, gained by the uniformly high quality of personnel and subcontractors used on the job, is maintained through careful quality control during the entire project. And while the main promotional efforts are aimed at owners, architects, and engineers, the firm also projects its image of quality to the public in general.

Case 3: Plumbing and Heating Contractor

This firm is typical in origin of many firms in the plumbing and heating field, as well as in many of the other special building trades. The firm was established by a journeyman plumber who with the help of his son managed to gross some $30,000 annually; he had neither the financial know-how nor the inclination to do more than earn a comfortable living for his family. When the father died, the son—a more aggressive businessman than his father—took over and moved into the commercial and institutional fields. Five years later, the firm grossed more than $3,000,000 with a profit margin of some 5 percent resulting in net profits of almost five times what the previous gross volume had been. Several factors were important in keeping profits in line with volume. First, a great deal of care was exercised in selecting jobs that, because of their size, location, or unusual characteristics, would be least likely to attract a large number of bidders. Second, each job to be bid was carefully examined for possible cost savings; on one job, for example, it was found that a different union (the United Mine workers, in this case) could be used, at a lower wage rate than the union used by his competitors, who failed even to consider such a possibility. While windfalls like this were infrequent, they paid off handsomely. For when unusual cost savings were found, they were *not* passed on directly to client through a lower bid, as is common practice among contractors; the jobs were bid essentially the same as before, giving the effect of a much higher profit margin with the same level of bid, and the same chance of being low bidder. Finally, office expenses have been kept to a minimum with a full-time office staff of only three persons plus a few outstanding supervisory personnel on the permanent payroll. This overall combination of lower costs and comparable prices, along with a high proportion of successful bids resulting from careful job selection, has resulted in one of the most prosperous contracting businesses of its kind.

Case 4: Mechanical Contractor

This firm, specializing in commercial air conditioning, was founded by a man who knew relatively little about the technical aspects of air conditioning system design and construction, but who was unexcelled as a salesman and probably unequaled by any contractor in terms of net profit on investment. Having identified the air conditioning field as an outstanding growth industry following his release from the Navy in 1945, this man worked the next five years as a salesman for several large air conditioning contractors. He considered this an excellent apprenticeship in the field, and after five years was able to buy out a small air conditioning—sheet metal firm that was suffering financial difficulties. He began his new business by hiring the best mechanical design engineer he could find (at the best salary the engineer could find) and a top serviceman for his shop. He then proceeded to sell a complete air conditioning package for commercial installations—design, construction, and service. He seldom bid a job, but negotiated work on the basis of his firm's extra capabilities for design and service. The quality of his work has never been questioned, either by clients or competitors. His annual volume is lower than most of his competitors, but his annual profits are probably double those of most firms having five times his volume. During the past six years he has averaged net profits of more than 100 percent on his invested capital. And he insists that any contractor can do likewise simply be selling his know-how and service instead of just man-hours and pipe; for everyone, he says, knows "the price of expertness always comes high."

Summary

To be successful a firm must analyze its present capabilities, establish realistic objectives, and develop workable plans to accomplish these objectives. While management is responsible for a number of different functions, all are concerned generally with using the firm's physical and financial resources to maximum advantage. Although there can be no absolute rules for successful contracting, certain general principles judiciously applied can almost certainly improve the firm's profitability.

6

Establishing the Company's Objectives

You seldom get what you go after unless you know in advance what you want. Indecision has often given an advantage to the other fellow because he did his thinking beforehand.

MAURICE SWITZER

The contractor who persistently directs his efforts toward achieving an honest, realistic objective will ultimately realize his objective. His frequent failure to do so merely confirms the truth of this statement—either the objectives were not realistic or he lacked the necessary determination and stamina in pursuing them.

To be useful the firm's objectives must be both realistic and attainable, based on unbiased estimates of what the firm can reasonably hope to accomplish in view of the overall market situation. Objectives are thus distinguished from hopes, which are based instead on vague, intangible optimism. Emotionally uncontrolled optimism provides a weak foundation for effective business planning.

Establishing realistic objectives involves a critical appraisal of the firm's capabilities, of the market in general, and of the firm's role in the overall market picture. In formulating realistic and attainable goals the firm will be forced to answer these broad, searching questions about itself:

1. What are its present capabilities and competitive strengths and weaknesses?
2. What needs to be done now to maintain or strengthen its position in existing markets?
3. What must be done to meet the competitive requirements of future markets?

57

The future conditions that will confront the firm as it works toward its objectives are largely the result of presently existing causes. While having realistic goals cannot eliminate the risks of the future, objectives will help the firm to recognize what needs to be done now to take advantage of the time lag between current events and their future effects—essentially, to exploit the future based on a study of the past and an analysis of the present.

The Importance of Having Objectives

Success refers to the accomplishment of specified goals. A business firm, before it can be successful, must define its objectives in specific, measurable terms.

Effective business planning requires that appropriate objectives be established and pursued. Without well thought out, clearly defined goals, the contractor will know only where he has been—and not where he is going.

Good objectives will help the contractor in a number of ways:

1. Planning personnel requirements.
2. Improving equipment utilization.
3. Controlling inventories.
4. Allocating overhead costs.
5. Anticipating financial requirements.
6. Forecasting profits.
7. Increasing awareness of the firm's operations.
8. Guiding company pricing policies.

Personnel Requirements

Most construction workers can be hired on a job-by-job basis, for there is generally a pool of available labor large enough to fill whatever needs might develop. Foremen and other supervisory personnel, as well as estimators and office employees, must be retained on a full-time basis, however, if the top-quality people necessary to assure efficient operations are to be available when needed. By establishing specific, attainable goals, the firm can anticipate the requirements for permanent office and field personnel and can justify keeping them on the payroll.

Equipment Utilization

Idle equipment is expensive. By planning ahead the demands on major equipment items can be estimated, thus making possible the identification and allocation of their costs on a more realistic basis. Objectives, so long as the chances of attainment are good, are also invaluable in making economically sound decisions regarding the relative merits of equipment purchase, rental, or leasing.

Inventory Control

Having more capital tied up in inventories than is absolutely essential means that less working capital will be available for other necessary purposes. Forward planning can identify the need for both the amount and type of inventory to be carried, as well as for the most economical timing of purchases.

Overhead Cost Allocation

Overhead costs are easy enough to identify in total, but extremely difficult to spread among a company's various jobs. These costs can be realistically allocated to specific jobs only if a fairly accurate estimate of total sales volume can be made. If a company's annual volume can be projected within reasonably narrow limits, overhead costs can be charged to projects with almost the same accuracy as direct job costs.

Financial Requirements

Becoming financially overextended is a major cause of business failure. By having specific objectives toward which to work, the dangers of financial overextension can be minimized. In addition, by continually monitoring the company's progress toward these goals, any changes in financial requirements can be recognized early, perhaps resulting in substantial savings in interest charges and almost invariably bringing about better uilization of available capital.

Forecasting Profits

The firm's growth depends primarily on the ability of management to employ profits effectively. During prosperous times, increased capacity brought about by reinvesting profits in the business can lead to increased profits. During lean times, however, increasing the firm's capacity has little merit, and profits can be withdrawn and used elsewhere to better advantage.

Increasing Management Awareness

The thought processes involved in establishing and analyzing a firm's objectives result in a keener awareness of the firm's overall operations and potential on the part of management and supervisory personnel. Further, by assigning responsibilities to individuals for certain performance goals, the probability of their accomplishment is increased.

Pricing Policies

Different objectives can be achieved in different ways, and a primary consideration in most approaches is the firm's pricing philosophy. Increased sales, for example, can be achieved most easily by lowering prices. To increase profits, however, usually requires that price levels be raised. And if the objective is to earn a certain return on invested capital, the optimum price may be higher than if a specific return on sales is desired. In any event, objectives will largely determine the company's pricing policies.

Types of Objectives

Different firms are motivated by different purposes, and management's initial task in setting goals is to decide exactly what tasks are to be accomplished and when. Many managements are satisfied simply to "make a living"; others are primarily interested in providing employment or in offering high-quality workmanship. Most, however, are concerned with making a profit, and various forms of the profit objective are most frequently used as business goals. The most commonly pursued objectives of construction management are directed toward the following:

1. Sales volume.
2. Gross profit.
3. Percentage return on sales.
4. Percentage return on total assets.
5. Percentage return on investment.
6. Improvement of performance.
7. Share of market.
8. Quality of performance.
9. Employment security.
10. Continuity of operations.

Sales Volume

The sales volume objective, while a common goal of construction firms, is a dangerous one to pursue without carefully considering its effect on profits. For almost any sales volume can be achieved in competitive bidding simply by setting prices low enough. With sufficient working capital to support a profitless operation, a contractor could conceivably capture almost an entire market— that is, until his funds were exhausted. Sales volume as an objective in itself, then, is economically unsound and should be discouraged. However, while sales volume cannot reasonably be pursued just for its own sake, neither can it be ignored, for only through sales can profits be generated. Sales volume should be looked on, therefore, as a means to an end—the end being profit—rather than as an end in itself.

Gross Profit

Gross profit is a function of the contractor's total sales volume, his fixed or overhead costs structure, and his percentage markups on direct job costs. Gross profits, while of primary importance, do not make as practical an objective as do profits related to other aspects of the firm's operations, since they measure neither the firm's earning capacity nor the efficiency with which profits have been generated.

Return on Sales

The percentage return on sales, also referred to as the "profit margin," is a frequently used profit objective since it can be taken directly from the profit- and-loss statement and provides a good indication of how the company fared over the past year. An added advantage of using the return on sales as an objective is that both the profit and the sales are measured in dollars having the same value, whereas either the net worth or total assets have probably been acquired over a period of years at varying price levels. The main disadvantage of using return on sales as an objective is that this measure does not show directly how effectively the firm's capital has been employed. For example, one company might require twice the capital investment to attain the same return on sales as another company that puts its capital to better use:

	Company A	Company B
Total sales	$100,000	$100,000
Total invested capital	20,000	40,000
Net profit before taxes	5,000	5,000
As percent of sales	5.0%	5.0%
As percent of invested capital	25.0%	12.5%

In this example, to infer that the two companies were operating with com- parable efficiency would obviously be misleading, even though their profit-sales relationships were the same. Company A is actually using its capital twice as effectively as Company B as measured by its net profit per dollar of invested capital.

Return on Investment

The percentage return on invested capital is probably the best indicator of a company's financial performance and for that reason is usually the best way to set the company's objectives. The return on invested capital is a direct function of the profit return on sales and the rate of capital turnover (the sales/net worth ratio) and can be expressed as follows:

$$\frac{\text{profit}}{\text{sales}} \times \frac{\text{sales}}{\text{net worth}} = \frac{\text{profit}}{\text{net worth}}$$

A possible drawback to the use of net worth lies in the fact that the use of outside funds can exaggerate the firm's profitability, as long as these funds earn a return higher than the rate of interest paid on them. For example, by borrowing $100,000 at 10 percent interest and subsequently reinvesting it at only 12 percent, a firm with a net worth of $50,000 could increase the apparent rate of return on net worth by 4 percentage points by earning the additional $2000. If the firm was earning 16 percent on its $50,000 net worth, the borrowed capital would increase the return on net worth to 20 percent, even though these outside funds were earning only half the return shown by the rest of the firm's capital. Even so, this type of approach is often justified; whenever borrowed funds can be used to increase profits, their use may be warranted.

Return on Total Assets

The percentage return on total assets is another way of measuring profit in relation to investment, and as such shares some of the merits of the return on net worth method. The reason for using total assets instead of the owners' equity in the business is the theory that all funds, whether supplied internally or externally, should be employed profitably. The return on total assets has the virtues of providing an incentive to increase sales and reduce costs as well as to exercise extreme caution in adding assets which do not earn at a rate at least equal to the company's present return. This sometimes leads to overly cautious investment policies, however, since it tends to discourage the use of borrowed capital that might add to profits, even though earning at a lower rate than the firm's own capital.

Improving Performance

Many firms set their objectives simply "to do better than last year." This, in fact, may be an adequate objective in many cases so long as an attempt is made to define exactly which measures of performance are to be improved. However, if this is done there is no reason to be so vague, and the objectives should be expressed in more concrete terms.

Share of Market

Attaining a specified share of the total market is a popular objective in many industries (such as automobile manufacturing) but has little justification in contracting. As pointed out previously, a contractor can obtain virtually any volume of sales or any desired portion of the total construction market simply by lowering his prices sufficiently. As an objective, then, share of market is not feasible; it is, however, an important consideration in estimating a company's potential sales volume from industry totals, as a factor in determining the attainability of other worthwhile goals.

Quality of Performance

Some contractors desire nothing more than to be known for their high quality work. And in some fields, such as in light commerical, residential, and highly specialized work, there is a place for this type of operation. However, so much emphasis has been placed on the low-priced job in most construction programs that the contractor who stresses the quality of his work had best stay away from competitive bidding and concentrate on negotiated contracts.

Employment Security

There are many small contractors whose operations are aimed solely at providing employment for the owners and their men. In these operations there is little incentive to expand and little need for profits beyond what constitutes a secure and comfortable living. Essentially, the owner is working for journeyman wages, with an added feeling of independence from operating his own business.

Continuity of Operations

The sole purpose of many contracting firms is one of self preservation. Often, a father will want only to keep his business running until his son is capable of taking over the operation. This sometimes goes on for generation after generation, with the business neither gaining nor losing but merely perpetuating itself.

General Procedures in Establishing Objectives

The attainability of a company's objectives will be influenced both by the effectiveness of its operations and by the external environment in which it must operate. Objectives, therefore, must be geared to a combination of these elements—what the company can hope to achieve within its own market.

The future events that influence the company's success in attaining its objectives cannot be predicted with certainty. Fortunately, a high degree of certainty is not required, for most future events are brought about by present operating causes; anticipating the future effects of what is presently happening, or what has already happened, is usually adequate for planning purposes. The general procedure to be followed in defining the firm's potential in view of the overall market situation requires that four major factors be considered:

1. The general economy.
2. The industry.
3. The company's position in the industry.
4. Timing.

The General Economy

A study of the general economy provides a picture of the overall framework in which the industry and the firm must operate. The economy as a whole will determine to a large extent total expenditures for new construction. Abundant data are regularly compiled and are available from a number of government sources, industry groups, trade associations, private research organizations, and banks. From these data, sufficiently reliable estimates can usually be made of the total construction volume during the next year or so for a given area. Breakdowns can also be obtained showing the general types of construction expected, based on building permits, work authorized by public agencies, and work already on the drawing boards.

Industry Volume

From the general economic data, estimates can be made of the portion of the work falling within the contractor's own field or specialty. For example, a projection of 10,000 single-family housing units for the following year might result in $4 million worth of residential heating and sheet metal work, $1 million in air conditioning equipment sales and installation, and so on. Similarly, the estimated volume of commercial construction can be broken down into general construction work; electrical, mechanical, plumbing, and heating; and all its other component parts. Again, substantial amounts of data are, or should be, available from local trade associations and industry groups.

The Company's Position

The place to begin in estimating the company's share of its industry's total market is with the share of market that the company has already attained in the past. From this point management should attempt to determine why the company has achieved this share of market, based on an analysis of competitive factors, the company's strengths and weaknesses, sales relationships, financial condition, quality of workmanship, and other factors. Then, management must critically examine each of these factors to decide which competitive advantages can be maintained and which shortcomings can be overcome. Considering this information in conjunction with recent trends in the company's share-of-market position should give realistic and workable estimates of what the company can reasonably hope to achieve in the near future.

Timing

Objectives are of little value unless specific time limitations are set for their accomplishment. The type and size of contracting operation will dictate the actual time requirements used in establishing objectives and forecasting sales. Many successful companies work with monthly goals, forecasts, and performance reviews for the following one-year period, plus annual goals and forecasts for the next five years. Even the smallest firms should attempt to look at least one, and preferably two, years ahead.

Example of Profit Objective

Table 7 gives two examples of how objectives might be established for a medium-size contractor having a current net worth of $105,000 and annual sales of $525,000, with return on net worth used as the objective. Example A represents an increasing industry market, while example B shows a declining market. The firm's objectives are the same in both cases: to earn a 20 percent return on net worth next year, and to increase this return to 24 percent the following year.

TABLE 7. Establishing and Analyzing Profit Objectives

	Performance		Objectives	
	Last Year	This Year	Next Year	Two Years
Example A: Growing Industry Volume				
Total industry volume	$18,000,000	$20,000,000	$22,000,000	$24,000,000
Company sales volume	500,000	525,000	550,000	600,000
Percent of industry total	2.8%	2.6%	2.5%	2.5%
Net worth	$100,000	$105,000	$115,000	$125,000
Rate of capital turnover	5.0 times	5.0 times	4.8 times	4.8 times
Net profit before taxes	$15,000	$21,000	$23,000	$30,000
Percent of sales	3.0%	4.0%	4.2%	5.0%
Percent of net worth	15.0%	20.0%	20.0%	24.0%
Example B: Declining Industry Volume				
Total industry volume	$18,000,000	$20,000,000	$19,000,000	$17,000,000
Company sales volume	500,000	525,000	475,000	425,000
Percent of industry total	2.8%	2.6%	2.5%	2.5%
Net worth	$100,000	$105,000	$105,000	$105,000
Rate of capital turnover	5.0 times	5.0 times	4.5 times	4.0 times
Net profit before taxes	$15,000	$21,000	$21,000	$25,000
Percent of sales	3.0%	4.0%	4.4%	5.9%
Percent of net worth	15.0%	20.0%	20.0%	24.0%

The first step in either case is to summarize company and industry data for recent years. While only two years are shown in the examples, information covering at least the last five years would be useful.

The second step is to forecast the industry volume over the period for which objectives are to be set; in the example, a two-year period is used.

Next, the portion of this total market that the company can reasonably expect to capture should be estimated. Unless some specific competitive advantage is expected, it is dangerous to assume a much larger share of the market than has already been achieved during recent years. In the examples a slightly lower share (2.5 percent) has been projected.

The effect of the firm's objectives on its other financial and operating requirements can then be examined. In example A, a portion of the profits can be retained in the business to increase capacity in the face of an increasing market, thus raising both the net worth and the net profit target.

In example B, however, the declining market situation precludes any major investments in additional capacity, and the profits had best be withdrawn and used elsewhere; net worth, therefore, remains constant, thus setting the required net profit at a slightly lower level than in the first example. In the declining market, either profit margins (net profit as a percentage of sales) or the rate of capital turnover (the sales–net worth ratio) must be increased to maintain a given level of return on investment.

If possible, the net worth might even be reduced in example B in keeping with the lower anticipated sales volume; this will result both in increasing the capital turnover rate at any given sales volume and in lowering the necessary profit margin. If, for example, the net worth could be cut back to $80,000, the capital turnover rate for the same sales volume would rise to 5.9 times for next year, and to 5.4 times for the following year. The net profit requirement, meanwhile, would drop to $16,000 and $19,000, and the necessary profit margin would be reduced to 3.4 percent and 4.5 percent of sales, respectively.

Reducing the net worth assumes, of course, that the same share of market can be captured with the lower net worth; usually this is a reasonable assumption. Most companies operate at peak efficiency when they run "lean"—when investments are kept to a bare minimum. A few prosperous years have led to the eventual downfall of many businesses as they found themselves burdened by heavy fixed costs during periods of more intense competition.

Measuring Progress Toward Objectives

Setting the company's objectives involves making estimates of a large number of variable factors, and changes in any of these factors can materially affect the firm's chances of accomplishing its objectives. Provisions must therefore be made to review periodically the firm's progress toward its goals.

Developing a system for measuring performance, then, is a necessary step to be taken after management has firmly committed itself to pursuing specific goals. Preplanned checkpoints should be established for monitoring progress. Many companies have found monthly reviews to be of considerable value. Then, if the company's performance has not met expectation, the causes for variance can be determined and corrected before any permanent damage is incurred.

There are two general causes for a company's failure to meet its goals: internal and external. Internal causes mean that the company itself is at fault, and changes in operating procedures are necesary to improve performance. External causes, on the other hand, are a result of the environment in which the firm operates and may mean that changes in the firm's objectives will be necessary.

In both cases provisions should be made for taking prompt corrective action to bring either operations in line with objectives or objectives in line with operations whenever any broad discrepancies develop between planned and actual accomplishments. Only through constant monitoring can objectives exert the strong, constructive influence on the company's future that they should.

Summary

Objectives are necessary to a company in anticipating its future needs and in planning its future growth. Establishing realistic goals involves firm management commitments as to the desired achievements and to the timing for their accomplishment. The most common types of objectives are concerned with sales and profits, although some firms may be motivated by other, less tangible goals. Profit objectives can be measured in terms of their gross magnitude or, better yet, as a percentage return on sales, assets, or net worth; return on net worth is probably the best measure of performance. The attainabililty of a firm's objectives depends on three major factors: (1) the general economic conditions, which determine the amount and type of construction expenditures in an area; (2) the total industry volume for the markets in which the firm operates; and (3) the portion of this market that the firm can reasonably hope to obtain. After objectives have been established the firm's progress toward its goals must be periodically reviewed and revised if necessary to assure maximum benefits from the planning.

7

Problems with Profits

*i suppose the human race
is doing the best it can
but hells bells thats
only an explanation
its not an excuse.*

DON MARQUIS

What is collectively the construction industry's problem is individually the contractor's problem. The chief obstacle confronting the contractor in his quest for a profit is having to operate in the type of competitive environment described in the preceding chapters.

With costs rising faster then competition will allow prices to rise, and with volume increasing just rapidly enough to keep total profits at about a constant level, the contractor must move fast just to stay where he already is, which is usually none too comfortable a position to begin with.

The contractor's dilemma has been stated before, and will be stressed again many times: *If he bids high enough to make a profit he cannot get a job; and if he bids low enough to get a job, he will not make a profit.*

Or, his problem can be stated more poetically:

> He who bids his work too cheap
> Never gets the cash to keep.
> While he who bids his work too high
> Never gets the chance to try.

Four important relationships are discussed in this chapter, all from the standpoint of the individual contractor:

1. The price/cost relationship.
2. The price/volume relationship.
3. The price/profit relationship.
4. The profit/volume relationship.

The examples presented in this chapter are drawn from the actual experience of contractors in several different fields. As such, they represent *only* the experience of these *specific* contractors, and it would be dangerous to apply the results anywhere else. However, the experiences related here are believed to be at least typical of those encountered throughout the industry, and can be easily verified by an individual contractor simply by referring to his own records. He will almost certainly find that similar relationships exist in his own situation.

Factors Affecting Profits

The different factors that determine the profitability of a contractor's operations and their relationships to each other are illustrated in Figure 5.

A contractor's profit is directly determined by three factors—cost, price, and volume. Each of these three factors is in turn influenced by other considerations

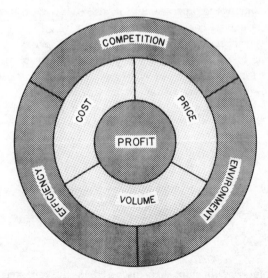

FIGURE 5.   Factors affecting profits.

that are largely beyond the contractor's control. But although he cannot *control* them, he can—and must—*react* to them to maintain a profitable business.

The *economic environment* determines how much work is available to the industry as a whole; *competition* determines how much of the available work can be obtained at a given price; and the *efficiency* of a contractor's operation determines how much it will cost to carry out the work that is obtained.

The total amount of new construction activity is influenced by a number of economic factors. Government policy, for example, has a strong effect on expenditures for highways, bridges, dams, sewer and water systems, land reclamation work, urban renewal programs, and other public works projects. General business and economic conditions, such as corporate profits and interest rates, exert their influence over the capital spending plans of corporate and other private clients.

Competitors cannot help but react to the economic environment, with their reactions most noticeably reflected in their markups on forthcoming jobs. A large volume brought about by prosperous business conditions usually results in a decrease in the intensity of competition and a higher overall level of markups. Leaner times caused by business recessions bring about more competition for the available work, with contractors' prices declining.

Most contractors are faced with about the same costs as their competitors in carrying out the work they get. Essentially the same labor, equipment, materials, and methods are available to all, and all encounter roughly the same costs of doing business. Overhead costs can (and should) be carefully watched and controlled; if managed efficiently, a large volume can provide a broader base over which fixed costs can be spread, thereby lowering the average cost per unit of work completed. Good planning, scheduling, supervision, and control are necessary to assure that costs are kept at least at a level comparable with competitors.

To make a profit the contractor must plan his operations on the basis of the business outlook, price his work on the basis of competitive factors, and carry out his work as efficiently as possible, preferably more efficiently than his competitors.

The Price/Cost Relationship

The percentage markup on direct costs determines the price/cost ratio, which is probably the most important single factor to the contractor in the profitable operation of his business. He can do little to gain a cost advantage over his competitors, yet his prices must be less than his competitors before he can get a job.

Figures 6 and 7 vividly illustrate, by means of the price/cost ratio, the problem confronting most contractors engaged in competitive bidding. These figures were compiled from analyses of about 1000 jobs involving several different classes of construction. They are composite figures representing the contracting field in general, but no single type of construction in particular.

Figure 6, a frequency distribution graph, shows the percentage of the total number of contracts awarded at various markups (price/cost ratios). Here, the "price" is the actual dollar amount for which the contract was awarded; "cost" is the estimated out-of-pocket direct cost (usually, just labor and materials) of performing the specified work.

Figure 6 indicates that more contracts were awarded to low bidders in the 100 to 110 percent range (0 to 10 percent markups) than in any other range. Just over 20 percent of the contract awards fell in this range. About as many, though, were let at prices ranging from 90 to 100 percent of cost, representing an out-of-pocket *loss* to the "successful" bidder of up to 10 percent of the job cost. About 3 percent of the contracts analyzed went at prices from 30 to 40 percent below cost. At the other extreme, 2 percent of the low bidders were able to realize markups of 60 to 70 percent above cost.

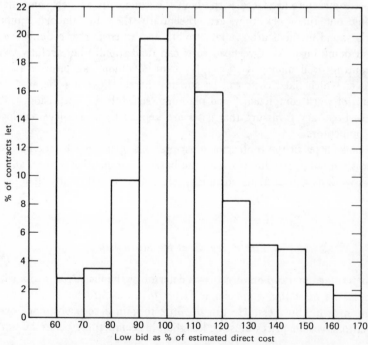

FIGURE 6. Composite frequency distribution of low bids.

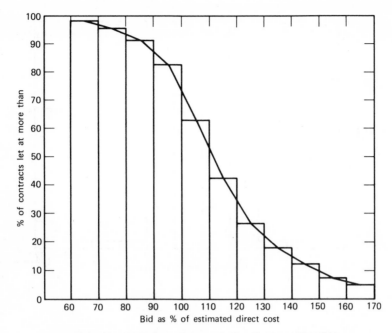

FIGURE 7. Cumulative frequency distribution of low bids.

Figure 7 plots the data from Figure 6 cumulatively, showing the percentage of total contracts let at or above a given markup. The median markup on these jobs was about 10 percent; one-half the jobs were let at better than a 10 percent markup, and one-half fell below that level. About 72 percent of these jobs were let at cost or above; 28 percent were let at below the estimated direct cost. Only one-fourth of the contracts allowed the "winning" contractor a 20 percent markup or more.

Figure 8 and 9 show the experience of individual contractors in four different fields of construction. Even though these figures refer only to specific operations, they do give some feel for the variety of conditions encountered in different types of construction activity.

As indicated in Figures 8 and 9, the bidding appears to be much "tighter" among prime contractors than among the subcontractors. For the general contractor (Figure 8*a*), nearly 95 percent of the contracts he bid on were let at prices in a range of 90 to 130 percent of cost. The building contractor (Figure 9*a*) had similar experience, with all contracts in his field going at a rate of 90 to 140 percent of cost.

In contrast, the painting contractor (Figure 8*b*) and the electrical contractor (Figure 9*b*) both encountered low bids falling anywhere from 40 percent below

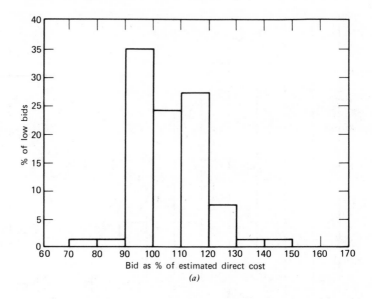

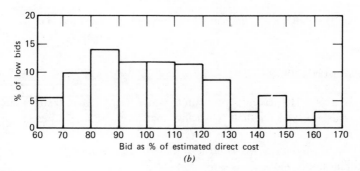

FIGURE 8. Frequency distribution of low bids in selected fields. (*a*) General contractor (highway-heavy); (*b*) painting contractor.

cost to 70 percent above cost. Apparently, more than 40 percent of the painting contracts were awarded at below-cost prices.

The Price/Volume Relationship

A contractor can get as much volume as he is willing to lose money on, if he will just bid low enough, but no amount of volume can ever make up for the lack of a profit.

The price/volume relationship is an important tool in business planning for the contractor. It shows him how much volume he can expect to get at a given price, and will be used extensively in establishing realistic objectives.

Figure 10 shows a typical price/volume graph for a general contractor. No attempt has been made to smooth the data or to make the curve conform to any theoretical shape. In fact, most of the data encountered by contractors will not fall in any theoretical manner, nor is there any reason to believe that it should behave according to any particular theory.

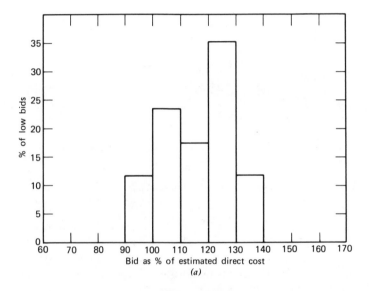

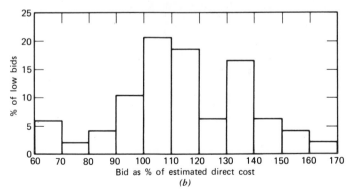

FIGURE 9. Fequency distribution of low bids in selected fields. (*a*) Building contractor; (*b*) electrical contractor.

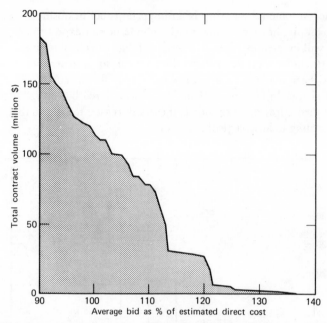

FIGURE 10.   The price/volume relationship (general contractor).

In the case illustrated in Figure 10, the contractor could generate about $180 million in volume by bidding all his work at 10 percent below cost (a price/cost ratio of 0.90). If he were to bid every job at cost, this volume would be reduced to $114 million. By increasing his markup to 10 percent, his volume would be reduced to $78 million; a 20 percent markup would result in $27 million worth of work; and a 30 percent markup would drop his volume to $3 million.

It is apparent in Figure 10 that the most critical area of concern for this particular contractor—the area where his volume is most sensitive to his markup, or where just a small increase in his markup will result in a large decrease in his volume—lies in the markup range between 5 and 15 percent. This range of markups represents a corresponding range in volume of from $114 million down to $30 million.

The data plotted on the price/volume graph are necessarily derived from historical data on past jobs. In the example just discussed, the preceding year's operations were used. Even though this approach implies that conditions over the next year will not differ significantly from those encountered during the past year, it still offers the contractor some real help in his operational planning.

The knowledge, for example, that to achieve a volume of $100 million would require markups of from 3 to 5 percent, or that a 25 percent markup would result in only a volume of $5 million, will certainly be useful as the contractor develops his strategic plans for the coming year.

The Price/Profit Relationship

While careless bidding can generate almost any volume desired, making a profit requires some careful thought.

The same data used in developing the price/volume relationship shown in Figure 10 can be used to compile a price/profit graph. Given a certain dollar volume achieved at a specified markup, the corresponding gross operating profit can be calculated as follows:

$$\text{gross operating profit} = \text{volume}\left(\frac{\text{markup}}{100 + \text{markup}}\right)$$

The "gross operating profit" referred to here represents the difference between the total amount received for the work and the *direct* cost (labor and materials) of performing the work. Thus, "profit" defined in this sense does not consider overhead expenses, indirect costs, or other deductions, and is not a true profit. It does, however, represent the total cash generated by a project and available to the contractor to cover these other costs.

The preceding arithmetic expression is obviously an important relationship, since it tells how much gross profit can be generated per unit of volume at any given markup. At a 10 percent markup, for example, the gross profit will amount to $10/(100 + 10)$, or $9.09 per $100 of sales, or 9.09 percent of whatever total dollar volume can be achieved at that markup. Similarly, a 20 percent markup will generate a gross margin amounting to 16.7 percent of the volume; a 100 percent markup on direct costs will result in a gross operating profit totaling one-half the dollar volume. Figure 11 shows the relationship between the percentage markup on direct costs and the resulting gross margin generated on the work obtained at that markup.

The price/profit relationship for the same general contractor whose operations were described in the preceding section is shown in Figure 12. It does not require a financial wizard to recognize that any bid below the out-of-pocket cost of doing a job will result in a loss on that job. Consequently, the price/ profit curve starts at cost (a zero percent markup) with a profit of zero. From there, the total profit will rise as the markup increases until the profit resulting from the higher markup no longer balances the decreasing volume brought about by the higher prices.

In this example, profits start at zero when jobs are bid at cost, and end at zero when a markup of more than 37 percent is applied. The one extreme gives

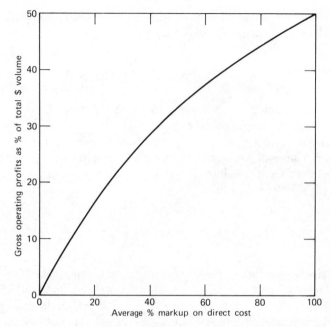

FIGURE 11.   Arithmetic relationship between markup and gross operating profit.

a large volume at zero markup, and the other extreme gives zero volume at a
high markup. As far as the total profit is concerned, it makes no difference
which of these two options is chosen. The author's personal preference would
be for no volume at a high price, which is a lot less trouble.

The price/profit relationship shown in Figure 12 is not the way a statistician
or economist likes to see data occur, but nevertheless it does represent the way
the data fell in this case. This contractor's total operating profits for the preceding year's operations hit a peak at a 10.5 percent markup, where $7.3 million
in profit would have resulted from a $77 million volume. As the markup
increases from that point, profits decrease to a level of about $3.7 million at a
13.5 percent markup, then increase to a second peak of $4.6 million at a 20
percent markup; then they fall off rapidly as shown in Figure 12 until they
reach zero at a 37 percent markup.

The price/profit chart shows this contractor that he could have realized his
maximum profit over the past year by applying an average markup of around
10 percent. If he preferred, he could have used a 20 percent markup instead,
and realized about one-half the gross profit on a volume only one-third as
large.

The Profit/Volume Relationship

In a conventional profit/volume chart, profit is normally expected to increase as volume increases. In some business, profits increase even at a faster rate than volume, as the increased volume leads to economies of scale.

A contractor's profit/volume chart, though, is more apt to look like Figure 13, because for a contractor to increase his volume it is usually necessary for him to cut his prices.

It can safely be concluded from this that construction contracting is not a conventional business and cannot be planned or operated like a conventional business.

Since a contractor's volume is directly related to his price or markup (as shown in Figure 10, the price/volume chart), the profit/volume chart will have the same shape and characteristics as the price/profit chart of Figure 12; only the scale along the horizontal axis will be different, reflecting the price/volume relationship.

In Figure 13, it is apparent that the maximum profit obtainable in this general contracting operation will occur at a volume of $77 million, where gross profits of $7.3 million will be realized. By reference to either Figure 10 or Figure 12, it can be seen that this point corresponds to a markup of about 10 percent.

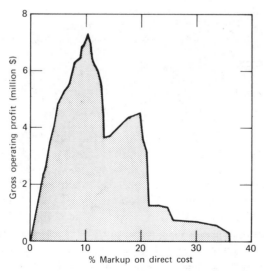

FIGURE 12. The price/profit relationship (general contractor).

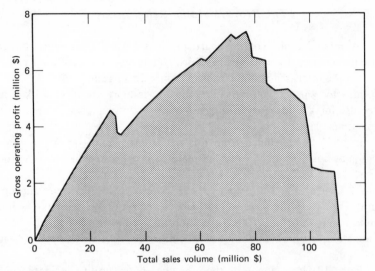

FIGURE 13.   The profit/volume relationship (general contractor).

The profit/volume curve starts at zero profit with zero volume and ends at zero profit with a volume of $110 million. As shown previously, to obtain a $110 million volume would require that the jobs be bid at cost.

While the profit/volume chart might be considered redundant in that the information presented on it is already available on the other charts, it still provides an important and useful reference for the contractor, showing the dollar volume required to generate different levels of profits.

Summary

The contractor's profits are determined by three factors: (1) price; (2) cost; and (3) volume. *Prices* are influenced primarily by competitive factors, while *costs* depend on the contractor's internal efficiency and controls, and *volume* is determined by external economic and business conditions. The *price/cost* relationship defines the prevailing level of competitive markups, and sets the upper limit on "what the market will bear" in pricing new work. The *price/volume* ratio measures the sensitivity of a contractor's volume to changes in his markups; it tells him how much volume he can obtain at a given markup level, a useful figure in his business planning. The *price/profit* relationship defines the amount of profit associated with whatever volume is achieved at a specified price, thereby indicating the general level of prices that should be aimed for in order to attain the highest total profits possible. The *profit/volume* relationship provides similar information, showing the amount of volume that would be required to generate the maximum amount of profits and revealing alternate ways of achieving any attainable level of profits.

8

Planning Your Profit Objectives

First, have a definite, clear, practical ideal—a goal, an objective. Second, have the necessary means to achieve your ends—wisdom, money, materials and methods. Third, adjust all your means to that end.

ARISTOTLE

The surest way to achieve maximum profits in competitive bidding would be to know every competitor's bid in advance, then bid each job just under the cheapest competitor's price. Whenever the lowest competitor's bid was below cost, of course, it would be best to skip the job.

Lacking complete prior knowledge of his competitors' prices, though, the contractor will have to settle for something less than maximum profits. When he is the low bidder, he will leave something "on the table," and in many cases his bid will be just a little too high to win the job, in which case he may feel that he has left it all on the table. How little less than maximum profits he has to settle for—how much he leaves on the table and how many jobs he should have gotten but did not—provides a direct and objective measure of exactly how effective his bidding strategy is.

With a reasonably good bidding strategy, a typical contractor should be able to make about half as much as he could if he knew all his competitors' bids in advance. At the same time, he should earn at least a 20 percent net return on his invested capital. In most cases, his markups should yield a net profit margin (or return on sales) of between 4 and 5 percent, although this figure will vary from field to field. His volume will be sufficient to "turn over" his capital from 5 to 10 times annually.

The contractor who attempts to run his business without having first established some realistic profit objectives is like a golfer who does not know in which direction or how far the hole is. To make the best possible score in either golf or business, it is necessary to have a known starting point, a specific goal or objective, and an understanding of what it takes to move from one to the other.

Profits and Profitability

While price, cost, and volume are the main factors that determine a contractor's profits, profits related to invested capital define the business's overall profitability. *Profits* measure the *amount* of money left over after all the bills are paid; *profitability* measures the *efficiency* with which the firm's capital was employed in earning the profits.

A number of important elements enter into the determination of price, cost, and volume, and ultimately, profits and profitability. These elements can be identified from the firm's two financial statements: the operating statement and the balance sheet.

Operating Accounts and the Profit Margin

Figure 14 shows the relationships between different items in a typical contractor's operating statement. Direct costs are made up of the out-of-pocket costs that can be attributed to a specific job, such as labor, materials, equipment operating costs (including fuel and maintenance), work that is subcontracted, and direct job overheads, such as payroll loadings and miscellaneous supplies and support materials.

Overhead costs include those items that are accrued with time, regardless of the amount of work handled; since they would be incurred whether or not a particular job were undertaken, they cannot logically be charged against any particular job. Some of the major accounts falling in this overhead cost category include officers' salaries, general office expenses, hired professional services such as legal and accounting, charges based on capital such as depreciation and property taxes, and other time-related items involved just with being in business.

Adding together the direct and indirect costs gives the total cost of doing business on the existing scale of operations. The ratio of indirect to direct costs gives the overhead rate, which may run anywhere from 10 to 50 percent for different contractors depending on their scale and type of operations and, even more important, on how they (or their accountants) classify the various expense items. A 15 to 30 percent overhead rate is typical for most contractors involved in engineered construction.

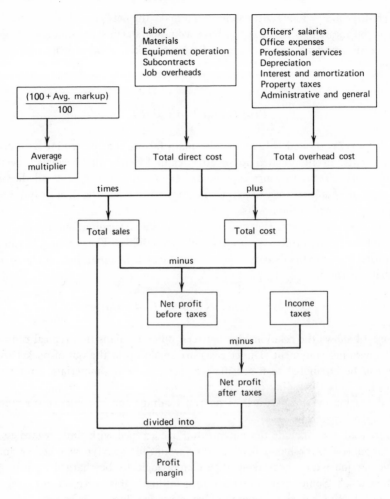

FIGURE 14.   Operating accounts and the profit margin.

Since the contractor's markup is generally based on his total estimated direct cost for a specific project, the direct cost plus the amount recovered through the markup on direct cost must equal the firm's total sales. In a break-even situation, the amount recovered through markups would equal the total overhead cost, total sales would equal total cost, and there would be no profit either before or after income taxes.

In a more normal situation, total sales will exceed total cost, with their difference being the firm's gross operating profit. Deducting income taxes from the net pretax profit will leave the net profit after taxes. This figure, divided by total sales and expressed as a percentage, is the profit/sales ratio, usually referred to as the profit margin. Profit margins typically run from 1.5 to 3 percent. Typically, however, contractors do not achieve the profit margins that they should.

Capital Accounts and the Capital Turnover Rate

While Figure 14 does identify many of the important factors that determine the contractor's profits, it does not indicate the efficiency with which the profits were generated, or the efficiency with which the firm's financial resources were employed. This requires examination of the company's capital accounts as reported on its balance sheet. Some of the major balance sheet items are shown in Figure 15.

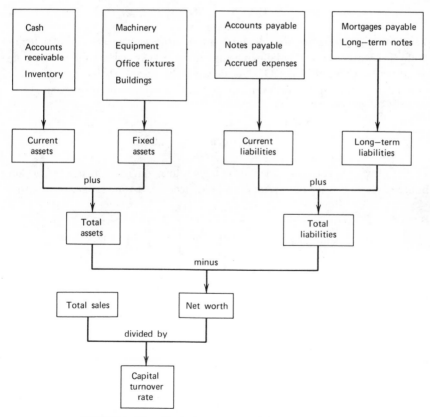

FIGURE 15.   Capital accounts and the turnover rate.

Four types of accounts are found in the balance sheet: current assets, which are readily convertible to cash; fixed assets, where capital is tied up over a relatively long time period; current liabilities, which are of a short-term enough nature that they must be covered by current assets; and long-term liabilities, where repayment schedules may extend over a period of years.

The firm's net worth—reflecting the owner's equity in the firm's assets—is simply the difference between total assets and total liabilities. The net worth represents the capital which must be profitably employed if the business is to be successful.

Dividing the firm's total annual sales by its net worth gives the capital turnover rate, a direct measure of how actively the capital is being used. Most contractors will turn over their capital from 4 to 8 times annually; in other words, their total sales volume will be from 4 to 8 times their net worth.

Return on Investment

Together, the profit margin and the capital turnover rate determine a firm's profit/net worth ratio, also known as its return on invested capital or return on investment (abbreviated as ROI). As shown in Figure 16, the net profit margin multiplied by the capital turnover rate gives the net return on investment:

$$\frac{\text{profit}}{\text{sales}} \times \frac{\text{sales}}{\text{investment}} = \frac{\text{profit}}{\text{investment}}$$

Profitability—expressed as the percentage return on investment—is the key figure here. The other two factors can be varied over a wide range, yet together

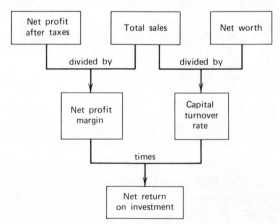

FIGURE 16. Determination of return on investment.

TABLE 8. Financial Characteristics of Various Types of Contractors

Type of Contractor	High	Typical	Low
General (buildings)			
Profit margin (%)	2.5	1.5	1.0
Capital turnover rate	12.0	8.0	5.0
Return on capital (%)	30.0	12.0	5.0
General (highway-heavy)			
Profit margin (†)	5.0	3.0	2.0
Capital turnover rate	6.0	4.0	2.5
Return on capital (%)	30.0	12.0	5.0
Mechanical			
Profit margin (%)	3.0	2.0	1.2
Capital turnover rate	10.0	6.0	4.2
Return on capital (%)	30.0	12.0	5.0
Electrical			
Profit margin (%)	4.0	2.4	1.4
Capital turnover rate	7.5	5.0	3.6
Return on capital (%)	30.0	12.0	5.0

yield the same overall result in terms of profitability. Just as in driving from one place to another, where several alternative routes may be available that all begin and end at the same place, the contractor can achieve a given net return on his invested capital in different ways.

Table 8 illustrates the ways in which contractors in different fields and operating in different competitive situations may still end up with operations showing comparable levels of profitability.

As shown in Table 8, building contractors generally have the lowest profit margins (profit/sales ratio) on their work. However, their low return on sales is compensated for by a more rapid rate of capital turnover. Much of the building contractor's work is subcontracted at fixed prices, thus reducing his own risk level and requiring substantially less fixed investment than would be required were he performing the work himself. In this case, the major portion of the total fixed investment required for building construction must be borne by the subcontractors.

General contractors engaged in engineered construction, on the other hand, typically have profit margins about double those of the building contractor. However, their investment in heavy equipment increases their capital requirements and decreases their capital turnover rate to just half of what the building contractor normally experiences. Still, a 1.5 percent profit margin and a turnover rate of 8 times, and a 3 percent profit margin accompanied by a turnover rate of 4 times, both work out to the same 12 percent return on net worth. Thus, either way, the contractor's capital can be considered as being employed with equal efficiency.

The major subcontractors—mechanical and electrical—usually fall between the building contractor and the heavy contractor, both in terms of their profit margins and capital turnover rates. Mechanical contractors, turning over their capital 6 times annually with a 2 percent profit margin, more nearly resemble the building contractors. Electrical contractors, with a 2.4 percent return on sales and a turnover rate of 5 times, earn their 12 percent return on capital more like the engineering constructors.

The figures shown in Table 8 are not the result of any statistical survey or tabulations, but they are based on experience with a number of firms in each field and are believed to be representative of typical and above and below average performance in each field.

In setting his profit objectives, the contractor will generally want as much profit as he can get. How much he *can* get, of course, may have little resemblance to how much he *should* get, or how much he is entitled to in view of his technical capabilities and the risks to which his business is subjected.

Certainly, a 20 percent net return on invested capital is not an exorbitant level of profitability in a business as risky as contract construction, and is a reasonable goal for the contractor to work toward. If a higher rate of return can be achieved, it should be sought; 20 percent is suggested simply because it seems to offer reasonable compensation for the contractor's entrepreneurial efforts, technical skills, and normal business risks. Also, 20 percent is usually about all he can get.

The Net Revenue Requirement

The *net revenue requirement* (NRR) is defined here as the total dollar amount above direct job costs that must be recovered through the markup in order to generate a specified net profit after taxes. Referring back to Figure 15 we see that the NRR must include provision for the following:

1. Total overhead costs.
2. Net profit before taxes.
3. Income taxes.

The contractor, through his markup on direct costs, must generate sufficient income to gain the desired net after-tax profit. This total amount received above direct job costs must pay all overhead costs, leaving a net profit before taxes that will enable him to pay his income taxes and still leave enough to satisfy his profit and profitability objectives.

The profitability objective is based on the balance sheet, but the profits must be earned on the income statement.

For example, consider a contractor having a net worth of $500,000 and overhead costs of $150,000 annually. His profitability objective is to earn 20 percent on his net worth, which translates into a profit objective of 0.20 × $500,000, or $100,000.

In his situation, net after-tax earnings of $100,000 would place him in a 50 percent income tax bracket, so his net pretax return would have to be $200,000. Adding the $150,000 in overhead costs to this $200,000 gives him a net revenue requirement of $350,000.

In order for this contractor to achieve his profitability objective, therefore, his operation must generate $350,000 above the total direct costs of performing the work. This $350,000 can be obtained only by providing an adequate percentage markup on an adequate volume of work.

But as was pointed out earlier, a given level of profitability can be achieved in different ways. Here, the $350,000 can be earned by applying a low markup on a large volume (such as a 3.5 percent markup on work estimated at $10,000,000), or by taking a high markup on a relatively small volume (35 percent on work costing $1,000,000, for example).

The amount of work needed to return any given amount is a function of the markup, and can be expressed as follows:

$$S = NRR\left(\frac{100 + M}{M}\right)$$

were S represents the sales volume required to generate the prescribed net revenue requirement (NRR), at a given average percentage markup, M, on estimated direct job costs. Or, the equation can be restated as

$$\frac{S}{NRR} = \left(\frac{100 + M}{M}\right)$$

thus giving the sales required per dollar of NRR at a given markup. This sales-to-NRR ratio is tabulated in Table 9 for various markups ranging from 1 to 100 percent; the relationship is illustrated graphically in Figure 17.

If this formula was applied to a contractor with a net revenue requirement of $350,000, then he would require the following sales to achieve his profitability objective:

- $12,015,000 at a 3 percent markup.
- $ 7,350,000 at a 5 percent markup.
- $ 3,850,000 at a 10 percent markup.

- $ 2,685,000 at a 15 percent markup.
- $ 2,100,000 at a 20 percent markup.
- $ 1,750,000 at a 25 percent markup.
- $ 1,225,000 at a 40 percent markup.

At a 100 percent markup this contractor would need only $700,000 of sales. Figure 18 shows the total sales required at various markups.

The problem facing this contractor becomes more apparent when expressed in this form. He may not be able to get $12,000,000 worth of work even if he bid all his work at cost, and he would certainly not be able to generate a $1,200,000 volume if his work were priced at 40 percent above cost.

The remaining question to be answered before he can develop his overall pricing policies is simply this: How much sales volume can be obtain at the various markups? To find the answer, he must examine next his price/volume relationships.

The Price/Volume Relationship

Figure 10 showed a typical price/volume graph for a general contractor and reviewed some of the implications of the contractor's price/volume structure.

Unfortunately, there is no such thing as a "typical" contractor. The relationships between percentage markups and the sales volume needed to reach a

TABLE 9.   **Sales Required per Dollar of Net Revenue Requirement at Various Markups**

Markup on Direct Cost (%)	Sales Required per Dollar of Net Revenue Requirement	Markup on Direct Cost (%)	Sales Required per Dollar of Net Revenue Requirement
1	$101.00	13	$8.69
2	51.00	14	8.14
3	34.33	15	7.67
4	26.00	20	6.00
5	21.00	25	5.00
6	17.67	30	4.33
7	15.29	35	3.86
8	13.50	40	3.50
9	12.11	45	3.22
10	11.00	50	3.00
11	10.09	75	2.33
12	9.33	100	2.00

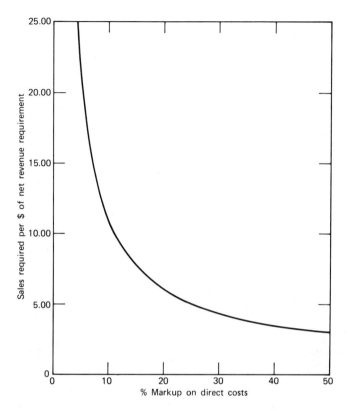

FIGURE 17. The sales–NRR markup relationship.

specified NRR are developed mathematically and apply universally to all types of contractors; but each contractor will have different price/volume characteristics. There is a no general shape for a contractor's price/volume curve that can be expressed in any meaningful mathematical terms.

Continuing with the example of the contractor having a $350,000 net revenue requirement, assume that, by examining his records on past jobs, he finds that he would have obtained the following sales volumes over the past year, had he bid all of his jobs at the specified markups:

- If he had bid every job at cost, he would have had a total sales volume of $8,500,000 from the jobs he bid last year.

- Had his markup been 5 percent on every job, his volume would have been $5,760,000.

- A 10 percent markup applied to all jobs would have resulted in a $4,320,000 volume.

- Increasing his markup to 15 percent across-the-board would have decreased his volume to $3,000,000.

- Going to a 20 percent markup would have dropped his volume to the $2,520,000 level.

- A 25 percent markup would have yielded total sales of $1,400,000.

- With a 30 percent markup, a $1,080,000 volume would have been realized.

- At a 35 percent markup, only $375,000 worth of work would have been obtained.

- And at a 40 percent markup, he would have had no jobs and no sales volume at all.

This information adequately defines the contractor's price/volume structure. In most cases, the data can be readily obtained from the contractor's own cost

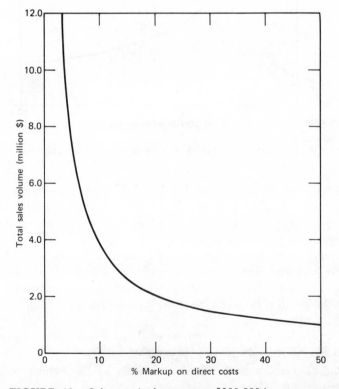

FIGURE 18.   Sales required to generate $350,000 in net revenue.

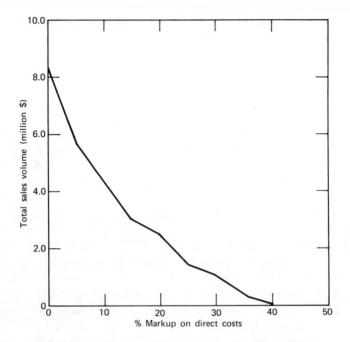

FIGURE 19.   Sales volume obtainable at various markups.

records and from the published tabulations of jobs bid. The price/volume chart for this particular operation is shown in Figure 19.

Figure 19 illustrates what has been said several times before. A contractor can achieve almost any volume he wishes, just by bidding low enough. If he is willing to work cheap, he will have plenty of opportunities to do just that. If he wants to place a high markup on his services, he can do so, but a relatively small construction market exists for high-priced work, and where competitive bidding is involved, the market is essentially nil. Still, the contractor should be aware of the existence of a choice, even if the choice is between near-catastrophic situations—in this case, between a high volume at a low markup or no volume at a high markup.

The next task is to find a compromise solution that will enable the contractor to realize his objectives.

Putting it all Together

Having determined the volume necessary to generate the desired NRR at various markups and the volume which could realistically be obtained at the corresponding markups, it remains for us only to combine this information. Table 10 shows the results.

Table 10 contains the same data as plotted graphically in Figure 18, which shows the volume needed at various markups, and in Figure 19, where the price/volume relationship is plotted. Superimposing the two figures gives the result shown in Figure 20.

In Figure 20 the shaded area defines the only area in which the firm's profitability objectives can be met. It is only within this area that sufficient volume can be obtained to satisfy the net revenue requirement. Outside the shaded area, there is either too little volume for the markup, or too little markup for the volume. In this example, the profitability objective can be met by applying markups ranging anywhere from about 7.5 percent to 22.5 percent. Lower markups provide insufficient profit margins; higher markups result in insufficient volume.

When the volume curve lies above the NRR curve, the vertical distance between the two curves represents the dollar amount by which achievements will exceed objectives. Anywhere in this area, the contractor will be earning more than his objective of a 20 percent return on net worth. At a 10 percent markup, for example, his volume would run $470,000 above what he actually needs to meet his profitability goal.

If, on the other hand, the volume curve falls below the NRR curve, the contractor is failing to achieve his profitability objective at that markup. Again, the dollar amount by which he misses his goal is the vertical distance between the two curves. At a 5 percent markup, his volume falls $1,590,000 short of what is needed; at 25 percent he is $350,000 short.

TABLE 10.   Sales Volume Obtainable and Needed at Various Markups

Markup on Direct Cost (%)	Volume Obtained at Specified Markup	Volume Needed to Generate $350,000 in Net Revenue	Difference
3	$6,800,000	$12,015,000	($5,215,000)
5	5,760,000	7,350,000	(1,590,000)
10	4,320,000	3,850,000	470,000
15	3,000,000	2,685,000	315,000
20	2,520,000	2,100,000	420,000
25	1,400,000	1,750,000	(350,000)
30	1,080,000	1,515,000	(435,000)
35	375,000	1,350,000	(975,000)
40	0	1,225,000	(1,225,000)

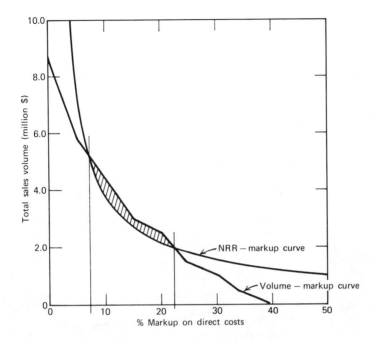

FIGURE 20.   Identifying the area where profitability objectives can be achieved.

Should the net revenue requirements be different from those assumed here, then the ranges of markups over which the NRR can be achieved will also be different. Figure 21 shows the effect of raising or lowering the NRR in this example by $50,000. A $400,000 NRR, assuming all other factors stay the same, would represent a 25 percent return on net worth to this contractor. Similarly, a $300,000 NRR would correspond to a 15 percent return on invested capital.

In Figure 21 it can be seen that a NRR of $300,000 can be realized at markups falling anywhere between 6 and 24.5 percent, while a $400,000 NRR can be obtained—just barely—between 10.5 percent and 21 percent. If the contractor's profitability objective were set at 30 percent (requiring a net revenue generation of $450,000), the objective could not be obtained under any conditions unless some of the factors were changed. This might be accomplished either by increasing volume or by reducing the revenue requirements. To increase volume, more jobs would have to be bid without bringing about a corresponding increase in either net worth or overhead costs. The net revenue requirement for the specified return on investment could be reduced by cutting down overhead costs or by getting rid of unneeded capital items, thereby lowering the firm's net worth.

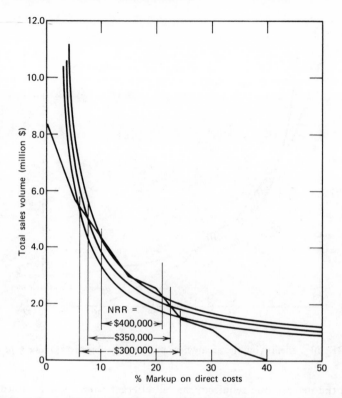

FIGURE 21.   Effect of NRR on markup ranges required to meet objectives.

Ratio Analysis

Analyzing a company's financial statements is an important step in judging management effectiveness. Management, too, can benefit greatly from the analysis, in spotting weaknesses and potential trouble spots more accurately and quickly than by other means, thereby providing a solid basis for management control.

The two major financial control instruments are the balance sheet and the profit-and-loss (or income) statement. The balance sheet shows what the company's resources are as of a specific date, while the profit-and-loss statement describes the company's operations over a period of time. Tables 11 and 12 are composite financial statements for contracting firms, developed from corporate income tax returns of more than 39,000 different firms representing all fields of construction. While the actual figures are of little significance, the relationships among these figures are valid and should be useful for purposes of comparison.

A company's current financial standing and past financial performance can be interpreted most conveniently when the relationships among various balance sheet and profit-and-loss statement items are expressed as ratios. The calculated ratios for a company can then be used to compare the company's performance with the average performance of other businesses of the same type. Similar comparisons can also be made among a company's own financial statements over a period of time, thus showing the firm's progress.

Several general rules stand out as being of particular importance in sound financial management and are worthy of special note:

1.   Avoid excessive investments in fixed assets, for insufficient working capital may be left, fixed charges will become burdensome, and the breakeven point will be unnecessarily high.

2.   Maintain working capital in correct proportion to sales to prevent the business from becoming financially overextended.

3.   Keep investments in inventories to a minimum, since having cash tied up in inventories may mean leaving too little cash on hand to cover current liabilities.

TABLE 11.   Balance Sheet for Typical Construction Firm

Assets		
Cash	$27,000	
Notes and accounts receivable	83,000	
Inventories	23,000	
Other current assets	9,000	
Total current assets		$142,000
Depreciable assets		35,000
Other assets		49,000
Total assets		$226,000
Liabilities and Capital		
Accounts payable	$57,000	
Short-term notes payable	17,000	
Other current liabilities	15,000	
Total current liabilities		$89,000
Other liabilities		47,000
Total liabilities		$136,000
Net worth		90,000
Total liabilities and net worth		$226,000

TABLE 12.   Income Statement for Typical Construction Firm

Gross sales		$500,000
Cost of operations	$407,000	
Expenses		
Compensation of officers	$16,000	
Rent	2,000	
Repairs	2,000	
Interest	2,000	
Property taxes	7,000	
Depreciation	9,000	
Employee benefits	2,000	
Other expenses	36,000	
Total expenses	$76,000	
Total costs		$483,000
Net income before income taxes		17,000
Income taxes		6,000
Net profit after taxes		$11,000

Significant Financial Ratios for Contractors

There are three basic kinds of financial ratios. The first group includes balance sheet ratios, showing the relationships among various balance sheet items. The second type of ratio, operating ratios, shows the relationships of expense accounts to income accounts on the income statement. The third group is composed of ratios showing the relationships between items in the income statement and items in the balance sheet. These three types of ratios provide management with the basic information needed for control over the company's costs and finances.

Many different ratios can be derived from the balance sheet and the income statement. In all, more than 30 different ratios are regularly compiled by various trade associations. Of these, 12 ratios have particular significance for contractors. These 12 financial ratios, and the average values of each for the 39,000 contracting firms represented in Figures 11 and 12, are as follows:

1.	Net profit to net sales	3.4 percent
2.	Net profit to net worth	18.9 percent
3.	Net profit to working capital	32.1 percent
4.	Net profit to total assets	7.5 percent
5.	Sales to net worth	5.6 times
6.	Sales to working capital	9.4 times

7.	Sales to fixed assets	6.0 times
8.	Current assets to current liabilities	1.6 times
9.	Cash to current liabilities	30.3 percent
10.	Fixed assets to net worth	93.4 percent
11.	Current liabilities to net worth	98.8 percent
12.	Total liabilities to net worth	1.5 times

Net Profit to Net Sales

The net profits on sales, or "profit margin," is an important yardstick in measuring profitability. It indicates to some extent a company's competitive strength, or it vulnerability to a decrease in either its sales volume or its profits. Usually, an increase in sales will widen the profit margin, since fixed costs need not rise in direct proportion to sales. Also for this reason, percentage changes in profits tend to increase and decline more rapidly than changes in sales.

Net Profit to Net Worth

The relationship between net profit and net worth is one of the most meaningful of all financial ratios and is often considered the best measure of profitability and efficient use of invested capital. In general, a profits-to-worth relationship of at least 10 percent is regarded as a necessary objective for providing funds for future growth. If the return on invested capital is much lower than 10 percent, the capital involved could probably be used elsewhere to better advantage.

Net Profit to Working Capital

Working capital represents the equity of owners in the current assets and is obtained by subtracting total current liabilities from total current assets. This equity, or margin, represents the "cushion" available to the business for carrying inventories and receivables, and for financing day-to-day operations. The ratio of net profits to working capital, expressed as a percentage, is useful in measuring the profitability of firms whose operating funds are provided largely through borrowings, or whose permanent capital is abnormally small in relation to the volume of sales.

Net Profit to Total Assets

The ratio of net profit to total assets is closely related to the net profit–net worth ratio. In this case, however, the theory is that the effectiveness of a company's operations should be analyzed in terms of all its assets, including capital provided by creditors as well as that furnished by investors, rather than in terms of equity interests only.

Sales to Net Worth

The sales–net worth ratio provides a means of determining the average turnover of capital during a given period; in other words, it shows how actively the firm's capital is being put to work. If capital is turned over too rapidly, liabilities build up excessively, as the amounts owed to creditors become a substutite for permanent capital. Conversely, if capital is turned over too slowly, funds become stagnant and profitability suffers.

Sales to Working Capital

This ratio measures the turnover of working capital. Most businesses require a margin of current assets over and above current liabilities to allow for cash paid out for stock and work-in-process inventory, and to carry ensuing receivables after work is performed and until the receivables are collected. This working capital turnover rate can highlight a financial problem if it is either very slow or very rapid. If the ratio between sales and working capital is too high, the tendency of the business is to owe too much, because it depends on credit granted by suppliers, banks, and others as a substitute for an adequate margin of current operating funds.

Sales to Fixed Assets

The ratio of sales to fixed assets is especially significant when compared with the same ratio from previous years, since such a comparison will show whether or not the funds used to increase productive capacity are being spent wisely. Any sizable investment in equipment or other fixed assets should lead to a larger sales volume. If corresponding sales increases have not resulted from these expenditures, poor utilization of assets—or poor investment policies—are indicated.

Current Assets to Current Liabilities

This ratio, also called the current ratio, is one of the most commonly used figures in analyzing balance sheets. It gives an indication of the margin of protection for short-term creditors. What constitutes a satisfactory current ratio varies considerably among different businesses. In general, the more liquid the current assets, the less margin that is needed to cover current liabilities. A current ratio of 2 to 1 is considered standard; a ratio of more than 5 to 1 is unnecessary and may, in fact, be a sign of weakness.

Cash to Current Liabilities

The ratio of cash and equivalent (marketable securities) to current liabilities, known also as the liquidity ratio, is an important supplement to the current ratio. The liquidity ratio indicates the immediate ability of a company to meet current obligations. A continual decline in this ratio may require the company to obtain additional capital, although it is seldom in itself a cause for major concern.

Fixed Assets to Net Worth

This ratio is used primarily to measure a firm's tendency to over-invest in fixed assets. A high ratio of fixed assets to net worth results in heavy depreciation and interest burdens, which in turn can lead to serious profit problems should sales difficulties be encountered. This ratio should seldom be allowed to exceed 100 percent.

Current Liabilities to Net Worth

The ratio of current liabilities to net worth, expressed as a percentage, provides a means of evaluating financial condition by comparing what is owed with what is owned. Whenever the relationship between current debt and net worth exceeds 80 percent, some financial weakness is indicated in that the business is overly dependent on its creditors.

Total Liabilities to Net Worth

When the total liabilities–net worth ratio exceeds 100 percent, the equity of creditors in the assets of the business is greater than the equity of the owners. Top-heavy liabilities make the firm extremely vulnerable to any unanticipated contingencies and severely limit management's flexibility.

Summary

The contractor's profits measure the amount of money left over after all bills are paid; his profits, related to the capital invested in the business, determine his profitability or the efficiency with which he put his capital resources to work in generating profits. The firm's objectives should be to use its capital as effectively as possible, attaining sufficient sales volume to turn over the capital as many times each year as is required to achieve a target percentage return on net worth. A typical contractor earns about 12 percent on his net worth, but should realize at least a 20 percent return on his capital. The total amount of money that must be made in excess of direct out-of-pocket job costs in order to achieve the desired level of profitability is known as the net revenue requirement (NRR), and is based on the amount of indirect costs, the target net return after taxes, and the income tax rate. The sales required to generate a given NRR is a function of the percentage markup on direct costs. Each dollar of NRR must be accompanied by $(100 + M)/M$ dollars of sales (where M is the average percent markup on direct costs) if objectives are to be met. The volume of sales that can be achieved at a given markup must be determined from bid tabulations on past jobs; the only generalization that can be made is that sales volume decreases as the markup increases. By plotting the NRR versus markup curve, then superimposing the sales volume versus markup curve, the contractor can quickly tell whether his objectives can be realized under existing competitive conditions; he can identify the range of markups required to achieve his profitability objectives; he can determine if his operation is being run as profitably as it might be; and he can decide what needs to be done to change his situation should changes be desirable.

9

Estimating and Controlling Costs

*When we mean to build, we first survey the plot, then
draw the model; And when we see the figure of the
house, then must we rate the cost of erection.*

WILLIAM SHAKESPEARE

Cost estimating and cost control are closely related subjects. In order to make accurate construction cost estimates in the office, actual costs of construction operations must be kept within fairly narrow limits in the field—this in spite of the many unpredictable and uncontrollable conditions that are constantly encountered.

The need for accurate cost estimates is obvious. If estimated costs are too high, the contractor will have little or no chance of getting a job in competition with others. If, on the other hand, estimated costs are too low, he stands an excellent chance of getting a job but no chance at all of making a profit. The risks of poor estimating, then, are all with the contractor: too high, no jobs; too low, no profits. While errors on the high side are costly, errors on the low side are disastrous.

Good, realistic estimates are important whether the contractor bids competitively for his work on fixed price or lump sum contracts, or obtains most of his work through negotiation with owners. Underestimating cost-plus work often results in hard feelings and a loss of owner confidence, sometimes a loss of reputation, and frequently a loss of future work.

And although inaccurate cost estimates represent one of the costliest aspects of the contractor's operation, estimates that are more accurate than necessary are also expensive. If a job does not offer reasonable prospects of making a profit, the job is not worth bidding; and if the job is not worth bidding, neither is it worth estimating. Fortunately, a variety of estimating techniques are available to the contractor; these techniques are valuable tools so long as their

103

limitations are recognized. The different types of estimates can be classified in three broad groups:

1. Order-of-magnitude estimates.
2. Semidetailed estimates.
3. Detailed estimates.

In preparing bids, there can be no substitute for complete, detailed takeoffs and pricing; however, shortcut estimating methods can give satisfactory levels of estimating accuracy in many specific situations in which the high cost of detailed estimates are not warranted.

Order-of-Magnitude Estimates

The order-of-magnitude estimate, sometimes called a factored estimate or a predesign estimate, provides a relatively low level of accuracy, perhaps varying by as much as 30 percent from the true cost. Accuracy can be considerably higher, however, if the contractor is thoroughly familiar with the type of work under consideration, and has been able to accumulate pertinent cost data on similar projects. The order-of-magnitude estimate can seldom be counted on to be within 15 percent of the actual cost, however, and should never be used as the basis for a bid.

Nevertheless, the order-of-magnitude estimate is a valuable tool for the contractor in quickly screening a large number of alternative projects to get a rough idea of what each job offers. The ability to cover a broad range of job possibilities in a short time frequently justifies sacrificing some precision in the shortcut method.

There are several different types of order-of-magnitude estimates. The most useful estimates are based on the following:

1. End-product units.
2. Ratios.
3. Physical dimensions.

End-Product Units

When sufficient experience has been gained with a particular type of work, charts can be prepared which relate end-product units to construction costs. Typical examples include the relationships between the cost of sewage treatment plants and the amount and type of sewage to be treated; electric generating plant construction costs and the plant capacity in kilowatts; apartment building construction costs and the number of apartment units; garage costs and the number of available parking spaces; and hospital costs for a given number of beds. The list could be practically endless, covering all types of applications where essentially the only difference between the proposed project and other completed projects is their size. Often the relationships between unit cost and project size will plot as a straight line on log-log graph paper, thereby giving the following mathematical relationship, where X is a constant factor for the particular type of project being estimated:

$$\left(\frac{\text{cost of project A}}{\text{cost of project B}}\right) = \left(\frac{\text{size of project A}}{\text{size of project B}}\right)^X$$

On large public housing projects, X equals approximately 0.8, when the number of rooms is used as a measure of project size. If a 300-room project was recently completed at a total cost of \$1,200,000, then the order-of-magnitude cost of a similar, 1000-room project could be estimated as follows:

$$\left(\frac{\text{cost of 1000-room project}}{\$1,200,000}\right) = \left(\frac{1000}{300}\right)^{0.8} = (3.33)^{0.8} = 2.6$$

The cost of the 1000-room project, then, will be 2.6 times the cost of the 300-room project, or about \$3,100,000. This type of mathematical relationship reflects "economy of scale"—where construction costs per unit of capacity decrease as the project size increases. The actual factor may range from as low as 0.4 to as high as 1.0, with the 1.0 factor reflecting a directly proportional relationship between size and cost.

Ratios

Some projects involve a single major equipment item, or several major items, which can be accurately priced. When this is the case, the costs associated with these major items, such as installation labor, appurtenances, materials, and so on, can be conveniently expressed as some percentage of the major item's cost. Ratio estimates are commonly used in the chemical and process plant construction field, where specialized equipment items often make up the major portion of the total project cost.

Physical Dimensions

Physical dimensions can be used to estimate the approximate cost of a variety of projects. Probably the most common types of estimates based on physical dimensions involve building construction, where fairly reliable cost estimates can be made in terms of either the building's square footage of floor space or its volume in cubic feet. Cross-country pipelines can often be estimated on a per lineal foot or per mile basis, and well-drilling can sometimes be estimated only on a per foot basis.

In all types of order-of-magnitude estimates, the accuracy to be gained is largely dependent on the estimator's judgment and experience; he must be able to visualize correctly the work as it will be done. The estimates must be made with care and used with caution; their limitations must be fully recognized. Any projects identified as being potentially favorable should be appraised in greater detail through more precise estimating methods.

Semidetailed Estimates

Semidetailed estimates, also called intermediate-grade estimates or budget estimates, should normally be accurate to within about 10 percent of the actual project cost, for this level of accuracy is usually considered adequate for making decisions regarding project feasibility—whether the owner will decide to proceed with construction. The "engineer's estimate" might be considered a semidetailed estimate in most cases. The accuracy of the semidetailed estimate is a direct function of the amount and quality of information available at the time the estimate is made.

More information is required for making semidetailed estimates than for making the order-of-magnitude estimate. Instead of using mathematical relationships between historical costs and estimated costs on proposed work, the contractor must consider the new project on its own. Actual quotations should be obtained on the major equipment and material items; some design data are necessary for making rough takeoffs; and approximate unit costs may be applied to the measured units.

The semidetailed estimates can be used to advantage by the contractor in several ways. Since its accuracy is generally sufficient to warrant the authorization of a project by an owner, the semidetailed estimate may be all that is needed on some negotiated cost-plus contracts, so long as the owner is aware of the accuracy limitations. This type of estimate is also useful in providing a rough check on detailed estimates obtained through more refined methods.

Detailed Estimates

Most contractors are intimately familiar with detailed estimates—the estimates used as a basis for making bids on construction work. These estimates, which should usually be accurate within 5 percent, are prepared from complete engineering specifications, drawings, and site surveys. Detailed estimates are time-consuming and expensive to prepare; consequently, they should be made only on jobs offering potentially good profit opportunities.

Estimating Procedure

In making good construction grade estimates, a large amount of detailed information is required. The first step in the estimating procedure is usually to examine the specifications that must be met. Then the project drawings—plans, elevations, schematics, and all the others—must be closely studied. Local labor availability and wage rates should be checked, quantity takeoffs on materials and equipment made, prices obtained on the specified materials and equipment, construction schedules drawn up, and subcontractors' prices called for on specialized phases of the work. Labor productivity is probably the most difficult of all items to estimate accurately, since many diverse factors affect the productivity of labor. Major considerations in making the labor estimates include the amount of shift and overtime work; expected weather conditions; the optimum ratio of supervisory personnel to workmen; the size of the project; and the effect of the engineer-inspector on labor productivity.

Unbalanced Estimates

Most contracts are let on a lump-sum basis. Some, however, call for unit-price bids where each item of work is priced separately, with the total contract price based on estimated quantities provided by the engineer. The contractor must be extremely careful in preparing estimates on these jobs. Unbalanced bidding—overpricing some items and underpricing others while keeping the overall total as accurate as possible—can often be used to advantage. The usual reason for unbalancing estimates is that monthly payments are made on the basis of measured progress, and a large part of the job cost can sometimes be recovered early if the early items are overpriced. A second use of unbalanced estimates occurs when the contractor has reason to believe that the quantities estimated by the engineer are not correct; if the contractor can overprice an item on which he thinks the quantity is understated, he may be able to get the job plus an additional profit for the added work at a relatively high price.

Estimating Accuracy

In every detailed estimate there are many opportunities for the contractor to make mistakes. These mistakes will make his estimate either too high or too low. And since errors of omission are more probable than errors of commission, mistakes will most likely result in estimates on the low side. The most common errors leading to low estimates are these:

- Omission of items.
- Undermeasurement of quantities.
- Underestimation of labor requirements.
- Failure to include sufficient allowance for overhead costs.

Types of Costs

The failure to recognize properly all elements of cost is one of the primary causes of trouble in the construction industry. Three types of costs must be included in every detailed estimate used as the basis for bidding:

1. Direct, or variable, costs.
2. Semivariable costs.
3. Fixed costs.

Direct Costs

Direct costs are those that can be attributed directly to a specific job, and should therefore be charged in their entirety against that job. Labor hired for a particular job and retained only for that job is an important direct cost; materials consumed during the job is another. Many labor-related costs that could be charged to jobs are instead charged as overhead costs; this practice, however, should be avoided as much as possible, since direct costs are much easier to work with, and to allocate to jobs, than are indirect costs.

Semivariable Costs

Semivariable costs are those which are only partly attributable to jobs and must be handled partly as overhead costs. Heavy construction equipment offers a good example of semivariable costs; while equipment operating costs may be charged directly to jobs, depreciation and taxes on the equipment continue regardless of the amount of work obtained.

Fixed Costs

Fixed, or overhead, costs are those related to time rather than to sales. Equipment depreciation and property taxes were already mentioned as overhead expenses; office rent and managerial salaries are also examples of major expense items that must continue even though sales levels drop.

Contingencies

After an estimate has been carefully prepared and checked, there should be some allowance for contingencies. This allowance should be based on the contractor's experience with respect to his actual performance on similar jobs. Most contractors are understandably reluctant to include a contingency allowance in their bid, correctly assuming that the job will be awarded to a competitor who did not include such an allowance. The purpose of the contingency allowance, however, is simply to make the estimate reflect the actual cost more accurately—and a contractor will never be penalized for having an accurate estimate.

The Cost of Estimating

The cost of preparing detailed estimates will vary widely according to the type of work being estimated. Jobs having a high proportion of their total costs in labor items will be more difficult and more expensive to estimate than will jobs in which materials make up most of the costs. Frequently, most of the cost of preparing an estimate falls in the overhead cost category, with only minimal out-of-pocket expenses incurred for travel, telephone, and site investigation. In such cases, the *incremental* cost of estimating may be insignificant.

Overall, estimating costs can be expected to range between 0.5 and 2 percent of the total project cost. A rule of thumb used by some contractors in estimating their costs on large jobs is as follows: total estimating cost = 0.005 × estimated direct materials cost + 0.015 × estimated direct labor cost. Using this formula, the total estimating cost on a $1,000,000 job having an equal split between materials and labor will be $10,000, or 1 percent of the total estimated project cost. The same size job, made up of $800,000 in materials and $200,000 in labor, would cost only $7,000 to estimate. A $200,000–$800,000 split between materials and labor would result in estimating costs of $13,000, or 1.3 percent of the total.

The amount and quality of data available on recent and similar jobs—especially regarding labor productivity—will determine to a great extent the costs of preparing estimates on present and future jobs. The estimate can be made as accurate as the contractor or his client is willing to pay for; the more accurate the estimate, the more time required for its preparation. Figure 22 shows the relative costs associated with the preparation of estimates having a specified level of accuracy, based on about a $1,000,000 project.

On relatively small jobs, estimating costs can be expected to run a higher percentage of the total job cost, while on exceptionally large jobs estimating costs might run lower. On jobs having a great deal of subcontract work, estimating costs would be lower for the general contractor; the subcontractor,

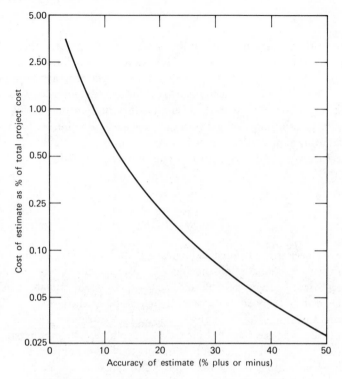

FIGURE 22.   Estimating the cost of cost estimating.

however, will still have the estimating costs, so estimating costs will necessarily be reflected somewhere in the total price of the job.

Every contractor whose operations require substantial expenditures for estimating should develop his own rules of thumb for estimating the costs of bid preparation. In many cases, he may find that his total expected profit on a job is substantially less than his costs of preparing detailed cost estimates for the work. By knowing his estimating costs, his chances of being low bidder, and the profit he can expect to make at the low bidder's price, he can eliminate many undesirable jobs and thereby eliminate many unnecessary costs.

Reducing Estimating Costs

The most effective, sure way of reducing the costs of estimating is simply to be more selective in choosing the jobs to be estimated and to reduce the total number of jobs bid. By taking a more selective approach and eliminating undesirable jobs in the preestimating stage, the contractor need not lose any profit-making opportunities; he is, instead, taking advantage of the same opportunities for profit, but at a lower overall cost. His total profits, then, should be correspondingly higher.

Several other ways of reducing the costs of estimating are also available to the cost-conscious contractor. One of the best is placing added emphasis on selling negotiated work in place of work obtained through competitive bidding. Another useful cost-cutting technique is to keep accurate historical cost records and use these data in developing shortcut methods of estimating. Even on some negotiated work, an intermediate level of estimating accuracy may be acceptable, and the need for extensive and detailed estimates can be eliminated.

Computer Estimating

Still another approach to cutting down the cost of estimating, this one a fairly recent development, involves using the electronic computer. Computers are especially valuable in estimating the costs of highly specialized types of work involving large quantities of well-defined units, especially when units of materials can be tied closely to labor requirements. Mechanical contractors, for example, have employed computers extensively in estimating the costs of piping installation; after materials takeoffs have been completed and current prices and wage rates have been fed into the computer program, estimates can be run off in only a few minutes. Thus, many man-hours normally required for tedious, routine arithmetic computations can be saved and perhaps devoted to more profitable tasks.

Requirements for Cost Control

Effective cost control is essential in every competitive business. A cost estimate is worthless if it fails to reflect the true costs of performing the specified work, and the best-managed, highest quality workmanship is wasted if the contractor is unable, through his bids, to get jobs on which his skills can be demonstrated. Since the cost control function provides the link between cost estimates and the actual construction project, a good cost control program is an absolute necessity.

The primary objective of the cost control function is to keep costs within the limits set by the cost estimates. This control is accomplished first by determining what costs should be, such as through standard cost data; then by identifying the areas where control is needed by comparing actual performance with the predetermined standards; by analyzing the causes of variances between actual and estimated or standard costs; and finally, by taking corrective action to bring costs back in line.

Judiciously applied, a good cost control program will help the contractor achieve better utilization of his manpower and equipment with a minimum of dealys or interference, and will provide him with the information needed to meet scheduled completion dates while staying within prescribed cost limits.

For any cost control system to function properly, a great deal of preliminary planning must go into every phase of the project; then this planning must be continued throughout the project's duration. The four essential elements in every effective cost control system are as follows:

1. Planning.
2. Estimating.
3. Scheduling.
4. Control.

Planning

The overall scope of the project must be clearly defined in great detail, considering the required quality and quantity of materials and workmanship involved. A project plan should be developed covering the work methods to be employed, the specific responsibilities to be delegated, and the persons responsible for each phase of the project.

Planning for each phase should rely heavily on the knowledge and experience of responsible individuals.

Estimating

The detailed cost estimates for the project should be made or compiled in terms of the same easily identifiable units as the construction schedule will subsequently be made in, so that actual progress will be both measurable and directly comparable with the estimates.

Scheduling

The project schedule defines the interrelationships between various parts of the project and specifies the time allocated for completion of each phase. A good construction schedule will give a clear picture of the exact role of every activity in the whole program and their effects on other activities. The element of time is the basis for the schedule, with the units of time associated directly with the same work functions used in the estimates.

Control

If work has been planned, estimated, and scheduled on a realistic basis, an appropriate system of measuring and reporting progress can quickly reveal any serious deviations from expected performance. The progress reporting system, then, must be tied closely to these other elements.

Measuring Progress

Objective measures are required that will indicate exactly where actual costs stand at any given time, relative to the costs estimated for the work that has been performed. The most common units of measurement used as indicators for cost control purposes are as follows:

1. Man-hours.
2. Actual costs.
3. Physical quantities.
4. Calendar days.

Man-Hours

By measuring the number of man-hours actually expended in relation to the amount of work accomplished, and then relating these to the original estimates, any major deviations from the estimates can be identified. Sometimes these relationships can best be expressed graphically by plotting the total cumulative number of man-hours against the elapsed time. Presumably, when half the man-hours scheduled for the project have been used, the project should be half completed. This technique can be quite effective on jobs where the degree of completion can be defined accurately in terms of end-product units. More often, however, man-hours should be broken down into specific categories—by crafts, for example—to give a better indication of the completion status of each part of the overall project. While such a procedure does not identify the causes of trouble that might be revealed, the fact that a problem exists will rapidly become apparent and further investigation into the causes can be made.

Actual Costs

The actual expenditures or commitments for labor, equipment, and materials can quickly reveal both the nature and magnitude of schedule and cost variations, and thereby indicate exactly where control action will be required to bring costs back in line. Actual costs are generally compared with standard costs for each major work function. Rapid, accurate tabulations of cost data from the field are especially important in this approach since some time lag is unavoidable in processing the information. Cost data should generally be collected daily, with summary reports compiled weekly.

Physical Quantities

Relating the physical quantities of work completed to the total project scope sometimes offers a convenient way of measuring actual progress against planned progress, although this approach lacks the direct tie-in with costs which is desirable. Not all types of construction projects lend themselves to easy measurement in terms of physical quantities, however; quantities are. sometimes difficult to define unless the project is broken down into hundreds of separate parts.

Calendar Days

The number of elapsed calendar days, when compared with the number of days budgeted for a particular operation, can reveal potential schedule-completion difficulties, thereby showing a need for added shifts or overtime work to meet deadlines. The main disadvantage of this method is that no indication of costs is given.

Reporting for Cost Control

If work has been planned, estimated, and scheduled on a realistic basis, a good progress reporting system can quickly reveal any serious cost deviations. Table 13 shows the types of information that can be included in a weekly summary report to management. Many contractors employ a computerized cost report instead, using essentially the same data and providing essentially the same type of information as shown in Table 13.

Table 13 uses all four units of measurments—calendar days, physical quantities, man-hours, and costs. Good results can be obtained, however, without necessarily using all four.

Each week's report should be carefully analyzed by management. Time lost can often be made up through added shifts or overtime work; money wasted during the early part of a project, however, can seldom be regained. Identifying potential trouble spots in their early stages from the standpoints both of time

TABLE 13. Weekly Summary Report on Installation of Concrete Footings

Week Ended	Estimated				Actual				Actual As Percentage of Estimated		
	Quantity	Percent Complete	Man-Hours	Cost ($)	Quantity	Percent Complete	Man-Hours	Cost ($)	Quantity	Man-Hours	Cost ($)
6/4	1,500	15.0	375	2,250	1,200	12.0	360	2,040	80.0	96.0	90.6
6/11	3,000	30.0	750	4,500	2,700	27.0	780	4,500	90.0	104.0	100.0
6/18	5,000	50.0	1,250	7,500	4,400	44.0	1,200	7,120	88.0	96.0	95.0
6/25	8,000	80.0	2,000	12,000							
7/2	10,000	100.0	2,500	15,000							

and money is therefore essential. Major problem areas can be easily recognized in the weekly reports.

First Weekly Report

The actual quantity of work performed during the first week is 80 percent of the amount estimated; labor costs, however, are already 96 percent of the estimated amount, and over 90 percent of the costs originally budgeted for the work have already been incurred. This indicates a dangerous situation in that costs are being accumulated at a more rapid rate than the work is being performed, as measured by predetermined standards. Man-hours, too, are getting out of hand, with nearly all of the budgeted man-hours being expended on only 80 percent of the work. Immediate corrective action is therefore called for.

Second Weekly Report

The situation has improved somewhat during the second week. But while all of the costs estimated for 30 percent of the total project have been incurred only 27 percent of the project has actually been completed. The apparent cause of this problem shows up in the man-hours expended—780 actual man-hours for the work completed, instead of 750 man-hours for the work planned.

Third Weekly Report

Here the situation continues to improve somewhat. Work is still behind schedule—44 percent complete instead of 50 percent complete—but the costs per unit of production have declined for the work completed during the week. If the rate of improvement shown since the first week can be continued, the project should be completed without too great an overrun. The work pace will need to be speeded up, however, for the project to be completed on schedule; this may create additional costs through overtime work or additional supervision.

Progress in this example has been measured in terms of total progress or cost to date against planned progress or cost for the same time period; comparisons could be made equally well in terms of costs per unit of work completed, or as progress or cost to date measured against the total amount budgeted for the entire project. The man-hour figures in this case are significant only in helping to point out the possible causes of cost variations.

The best method to use in measuring and reporting progress is largely a matter of personal preference and job characteristics; the method selected should be whichever one provides the most meaningful and usable information for management control.

Corrective Action

Different situations call for different types of corrective action; the cost report simply provides management with the information needed to identify where and when some degree of corrective action is called for.

Major cost items should always be emphasized, and given the best quality of supervision, for a large cost can usually be reduced by 10 percent just as easily as a small cost, and the benefits to the company are substantially greater.

Good management and supervision are the keys to effective cost control. Close control of costs in the field simply cannot be accomplished without adequate supervision, and poor cost control can almost always be traced directly to management shortcomings, whether in planning, scheduling, or supervising the work.

Many different types of cost control problems are apt to arise during a construction project, especially on large jobs where top-quality supervisors and foremen cannot exercise close control over all the different phases of the work. Most cost control problems fall into three broad categories:

1. Personnel.
2. Work methods.
3. Organization.

Personnel

Personnel problems resulting in cost variances may come from having too many, too few, unqualified, or inefficient personnel on the job. If the problem involves only the number of personnel available, crew sizes can be either increased or reduced. If necessary, additional personnel can be hired or shifted from other, less critical tasks; overtime work can be started; or parts of the work can be subcontracted. Additional or better supervision and perhaps more detailed definition of the work to be performed and the methods to be used in its performance can also increase labor efficiency and improve manpower utilization.

Work Methods

Work methods should be well established during the project's planning stages; if inefficiency is planned into the work, field personnel are sure to encounter control trouble. Nevertheless, inefficient work methods can often be corrected if they are detected in time by developing alternative methods of performing the work, by using more efficient equipment, by rescheduling the work, and by adding or improving supervision.

Organization

Some of the main causes of project overruns come before the actual field work even begins, during the estimating and planning stages. Oversights in properly defining the total project scope are common, estimating errors may be made, and labor rates and materials prices may increase. Failure to keep the costs of operations within the prescribed limits in cases like these, then, reflect more on the prescribed limits than on the operations. When the results of organization problems are encountered in the field, revisions in the construction schedule are often necessary to reflect the true conditions. Job plans must be revised, and extra efforts should be made to improve field operations by achieving maximum efficiency. The only truly effective way to correct flaws in project organization, however, is to make certain that the initial estimates, plans, and schedules are correct before field work begins.

Summary

Accurate cost estimating and effective cost control are essential to the contractor in making a profit. Cost estimates can be made at different levels of accuracy to serve different purposes. Order-of-magnitude estimates can be used to quickly screen a large number of jobs; semidetailed estimates are especially useful in making decisions regarding project feasibility; and detailed estimates serve as the basis for bidding jobs. For profitable operations, careful cost control is required in keeping actual costs within the limits set by estimates. Planning, estimating, and scheduling are all involved in cost control programs. Other necessary ingredients include a means of measuring progress and a system for reporting progress, so that the need for any corrective action can be quickly recognized and acted on by management.

10

Overhead and Fixed Costs

*Gain may be temporary and uncertain; but ever while
you live, expense is constant and certain.*

BENJAMIN FRANKLIN

Overhead costs are the most certain and easiest to estimate of all costs. Yet overhead costs cause more problems for more contractors than do all other types of costs. Although overhead costs may be easily identified and accurately estimated, they often cannot be assigned to specific projects. Direct costs, on the other hand, are difficult to estimate accurately but can be easily charged to a particular job.

Most overhead costs relate generally to the firm's overall operations rather than to specific jobs or work functions. A portion of overhead costs will be relatively stable; other overheads may vary according to the firm's capacity to do business, the amount of work bid, or the sales volume achieved. And still other overheads can be related to combination of two or more different factors.

Only in the very smallest of firms, where so few people are employed that all are needed full time, are overhead costs likely to be completely stable.

Overhead costs, unfortunately, are costs that some contractors feel can be reduced simply by reducing their prices on jobs. This thought is a dangerous illusion, for overhead costs can be reduced only in the office by management—and reducing the price bid for a job cuts profit, not overhead. Price cuts come off the top—right out of profits—while overheads go on forever.

Nature of Overhead Costs

To be engaged in contracting obviously requires more than hiring labor, buying materials, and obtaining equipment. The contractor must also provide the facilities, however small, for first getting the work, then for organizing, planning, and performing it. Being in business at all, then, entails some costs, with certain costs being incurred simply because the contractor is equipped to do business. These costs are referred to as overhead costs.

Overhead costs are largely independent of the amount of work actually done, or the total sales volume actually achieved. Overhead costs are relatively unaffected by outside influences, but are instead caused internally. The amount of overhead cost incurred is determined by the decisions of management; these costs are then accumulated with time, according to the obligations that management has established.

Overhead costs may be defined generally as all costs incurred by the contractor that cannot be atrributed directly to specific functions; these usually include all costs other than direct labor, materials, and equipment. Overhead costs fall into two categories:

1. Job overheads or indirect expenses.
2. General overheads.

Job Overheads or Indirect Expenses

Job overheads are the indirect expenses that vary with, and are caused directly by, individual jobs but are not directly chargeable against any specific phase of the work. These job overheads typically include such items as welfare fund payments, apprentice training, social security, workmen's compensation, unemployment taxes, miscellaneous payroll-related expenses, surety bonds, direct supervision, building permits, tool and equipment expenses, temporary buildings and enclosures, sanitary facilities, utilities, sales taxes, and many others.

Job overheads can be handled in different ways. Many contractors add job overheads to their estimates as some percentage of the estimated direct job costs. For contractors who perform essentially the same type of work all the time, and who maintain a stable work load, this method is generally satisfactory.

However, most contractors are confronted by varying work loads and different types of jobs, and a percentage add-on simply adds another element of uncertainty to the job. As long as costs can be identified and atrributed directly to certain jobs, these job overheads should be estimated with the same care and accuracy as the other direct job costs and included as such in the bid.

General Overheads

General overhead expenses are those which cannot be charged directly to a single job. These include the costs of maintaining an office or shop, such as rent, telephone and other utilities, property taxes, interest, insurance, salaries of office personnel (such as secretaries, bookkeepers, and estimators), association dues, sales and promotion expenses, office supplies, depreciation, and management salaries. Most of these general overhead costs must be paid regardless of the amount of work done or contracts received, although their magnitude may vary somewhat with the amount of business done or with the number and size of the contracts. Such expenses are continuous and can be controlled only through management's decisions to curtail them.

General overhead costs can be either fixed or variable in nature. Often even the variable general overheads are fixed within certain limits. Estimating expenses, for example, could be classified as a variable or semivariable overhead cost. A single estimator may be able to handle up to $1,000,000 in work annually; but to bid more work would require the addition of another estimator, which would then stabilize the estimating costs for perhaps another million dollars in estimating volume. Thus the estimating costs would not vary directly with the amount of work bid, but would instead be fixed within certain ranges, rising to a higher level as the estimating work load increased beyond the capacity of the available staff. And while estimating costs must be recovered, they cannot be charged to the jobs that were estimated—the total of all estimating costs must instead be regained through the relatively few jobs that are obtained.

The procedure for estimating the amount of fixed and variable overhead costs is simple and straightforward. The first step is to estimate the firm's total costs at zero sales volume; presumably, this amount would include most of the "time" charges—rent, utilities, property taxes, and so on. Then, the total overhead costs at the firm's expected sales volume would be estimated; these costs will include, in addition to the time charges, a portion of the costs of supervision, estimating, and other expenses that would be some function of the amount of work obtained, which should in turn be realted to the firm's capacity and the amount of work bid. The difference between the total costs at zero volume and the total costs at the expected volume, then, represents the variable elements of cost. The resulting figure can be divided by the volume to give an approximation of the variable overhead costs per dollar of sales.

In general, the fixed elements of overhead involve the resources for obtaining, or trying to obtain, jobs. The variable or semivariable elements are related to the actual amount of work that the firm either gets or tries to get.

Management Salaries as Overhead

Large firms nearly always include owners' salaries as an overhead expense item. However, many small contractors extract no salary for themselves, but instead draw on the firm's "profits" for their living expenses. This practice is somewhat misleading when the firm's financial position is appraised, since, in effect, the owner is contributing his time free of charge, thereby showing increased profits for the firm. In situations where the owner's salary or drawings are not looked on as a legitimate part of the cost of doing business, his living expenses must, nevertheless, be recognized as very real and necessary expense items and must somehow be reflected in his contract prices.

Depreciation

With the possible exception of management salaries, depreciation is the largest single item of general overhead expense for most contractors. And for some contractors—particulary those engaged in highway and heavy work—depreciation on equipment may well be the most important cost item.

Depreciation refers to the gradual exhaustion or wearing out of property used in the business, consisting of wear and tear, decay or decline from natural causes, and various forms of obsolescence. Depreciation refers only to the wearing out of equipment or property being used and is therefore based on original cost rather than on replacement cost.

Some distinction should be made between the accounting concept of depreciation and depreciation as it is used in estimating costs. For accounting purposes, a method of computing depreciation should usually be selected that will result in the greatest possible tax benefit; but in cost estimating, a method should be selected that will reflect most accurately the actual loss of value or utility of the property.

While the accounting concepts of depreciation are necessary from a tax standpoint, the actual depreciation costs of equipment are not necessarily related to the firm's accounting system. They are, instead, more closely related to the amount and severity of equipment usage. Probably the most realistic way of estimating the actual depreciation cost on equipment, therefore, is on a per-hour-of-use basis. By taking the total initial cost of the equipment and dividing by the estimated number of operating hours during the equipment's useful service life, a depreciation cost per hour of use can be established. This, in effect, amounts to a direct job cost chargeable directly to whatever jobs the equipment is to be used on, and can be handled as a direct cost rather than as an overhead item.

Contractors involved in exceptionally large and relatively long duration projects (such as dams) sometimes purchase new equipment for a specific job, sell the equipment at the end of the job, and charge the dollar difference between the purchase and selling price to depreciation. This method undoubtedly provides the most accurate and realistic measure of depreciation but is necessarily limited in its application for contractors in other lines.

These relatively direct methods of figuring depreciation can be used also to justify accelerated write-offs for tax purposes. Usually, however, one of these three conventional accounting methods will be used in tax calculations:

1. Straight-line method.
2. Declining balance method.
3. Sum-of-the-digits method.

Straight-Line Depreciation

The straight-line method is the simplest, but usually the least desirable, way to compute depreciation. Straight-line depreciation assumes depreciation to be equal for each year of the asset's life. Annual depreciation is found be deducting the asset's estimated salvage value from its original cost, then dividing the resulting figure by the estimated service life in years. An asset with a five-year life and no salvage value would therefore be depreciated at a rate of 20 percent annually. This rate gives a good measure of actual wearing out of equipment if annual usage is constant, but does not reflect the equipment's true loss in value, which would be substantially greater during the early years than indicated by straight-line depreciation.

Declining-Balance Depreciation

The declining balance method of computing depreciation uses a fixed percentage of the undepreciated portion of the asset's original cost, with the percentage generally set at twice the straight-line rate. For an asset with a five-year life and no salvage value, the first year's depreciation would be 40 percent of the original cost (twice the straight-line rate of 20 percent); each succeeding year would be 40 percent of whatever amount was left undepreciated.

Sum-of-the-Digits Depreciation

This method is similar to the declining balance method in that a relatively large portion of the capital investment can be recovered through depreciation charges during the early years of the asset's life. The allowable depreciation charge is computed by first adding together the digits representing each year of the asset's estimated service life. For a five-year life, the digits 1, 2, 3, 4, and 5 would be added together, totaling 15. The allowable first-year depreciation would be $5/15$ of the total initial cost; the second year, $4/15$ of the total could be written off; the third year, $3/15$; the fourth year, $2/15$; and in the final year, the remaining $1/15$ would be charged.

Choosing a Depreciation Method

Choice of the best method of accounting for depreciation of a firm's equipment and other assets should be left to competent accountants and experts on tax matters because many complex factors are involved. Additional depreciation allowances can sometimes be obtained, investment credits may apply, and proper timing or switching from one method to another can result in substantial tax savings. In general, depreciation methods permitting heavy write-offs early in the asset's life are preferable, since present savings are worth more than future savings. The final decision, however, should be made only after thoroughly examining the effect of each method on the firm's overall profitability.

Distributing Overhead Costs

Although the identification of general overhead costs is a relatively simple matter, their allocation is one of the most perplexing problems facing the contractor. Unlike direct job costs and job overheads—which are sometimes difficult to estimate accurately but which are easily identifiable in connection with specific jobs—general overheads are easy to estimate in total, but difficult to distribute equitably among different projects. The cause of this problem is that only a small portion of total overhead costs can be atrributed to each job, and the exact amount cannot be determined unless the total volume of work is known. Overheads, therefore, must be spread over an estimated but unknown amount of work.

Once the total amount of overhead expense has been determined, the main problem is to decide how much should be added to direct job costs so that the entire amount of overhead can be recovered. The contractor must be careful not to add too much, or he will be unable to get the necessary volume; should he add too little, however, he will be unable to recover the costs regardless of his volume. There is no one best way to allocate overheads to jobs; there are several different ways that are often used with satisfactory results. The most commonly used methods of distributing overheads are as follows:

1. As a percentage of total estimated direct costs.
2. As a variable percentage of estimated labor and materials costs.

Overheads as a Percentage of Total Direct Costs

The procedure followed by most contractors in allocating their overhead costs and profits is to base their estimates on the previous year's sales volume. The difference between total sales and total direct job costs for the preceding year, conveniently expressed as a percentage of sales, is used to determine the markup on work carried out during the current year.

This markup must first be converted from a percentage of sales to a percentage of cost. Then, the amount to be included for overhead on any specific project can be found by applying this percentage to the estimated direct job costs. Using this method, if volume during the current year increases over the past year, then profits should be higher. Conversely, if volume decreases, profits will decline.

The relationship between the markup as a percentage of sales and as a percentage of cost can be expressed mathematically as:

$$P_c = \frac{P_s}{100 - P_s}$$

where P_s represents the markup as a percentage of sales and P_c is the corresponding markup expressed as a percentage of direct job cost. For example, overheads totaling 20 percent of sales are found to amount to 20/(100–20), or 25 percent of direct costs. The mistake of figuring overhead as a percentage of sales, then adding that percentage to direct costs to arrive at a price, is not uncommon among contractors, and such a mistake can prove fatal when already thin profit margins are cut even further.

Overheads as a Variable Markup on Labor and Materials

There is a growing tendency among contractors to apply different percentage markups to labor and materials. There is certainly some economic justification for such a practice, since overheads which are directly attributable to labor and equipment will run much higher than those on materials and subcontracted work. Firm prices can usually be obtained on materials and subcontracted work, while the administrative expenses associated with these items will be relatively low. Also, high-labor jobs are much more risky than other jobs, and as such they require higher markups to compensate for the additional risks.

As an illustration of the effect of a variable markup on estimated job costs, assume that 30 percent is to be added to estimated direct labor and equipment costs, and 10 percent to the estimated cost of direct materials and subcontracted work. Table 14 shows the effect of this variable markup on three different jobs having the same total direct costs, and compares the resulting prices with those obtained by adding a fixed 20 percent overhead allowance to the total estimated cost of each job.

TABLE 14.   Effect of Varying Overhead Markups

		Total Overhead Allowance	
		Fixed 20% Markup on Total Estimated Cost	30% Markup on Labor and Equipment; 10% on Materials and Other Work
Job A			
Labor and equipment	20,000	$4,000	$6,000
Materials and other	80,000	16,000	8,000
Total direct cost	$100,000	$20,000	$14,000
Markup as percentage of total		20.0	14.0
Job B			
Labor and equipment	$50,000	$10,000	$15,000
Materials and other	50,000	10,000	5,000
Total direct cost	$100,000	$20,000	$20,000
Markup as percentage of total		20.0	20.0
Job C			
Labor and equipment	$80,000	$16,000	$24,000
Materials and other	20,000	4,000	2,000
Total direct cost	$100,000	$20,000	$26,000
Markup as percentage of total		20.0	26.0

Depending on the breakdown of direct costs between labor and materials, the effective overall markup ranges from 14 percent on job A up to 26 percent on job C. Only when costs are evenly divided between labor and materials will the results be the same.

The net result of using these variable markup rates, then, will be to price high-labor jobs higher and to price low-labor jobs lower. In effect, what can happen is that markups may be so high on high-labor jobs that there will be little chance of getting the jobs; and that markups will be so low on low-labor jobs that too great a volume will be required to recover the overhead costs.

Regardless of the method used to allocate overheads, however, the main objective remains the same: to recover a certain amount of money over a certain length of time. If a firm can obtain sufficient volume to accomplish this objective by means of a variable markup, the variable markup method is perfectly acceptable.

Overhead Cost Control

The basic problem in overhead cost control is not so much to eliminate or reduce overhead costs, but rather to be certain that the markups applied to direct costs are adequate, and the volume sufficient, to insure recovery of all overhead costs.

If overheads are not being recovered through the job markups, the contractor must find a way to do one of the following:

1. Increase his volume of work without increasing his overhead costs.
2. Reduce his overhead costs without reducing his work volume by eliminating expenses that do not contribute their share to income.
3. Raise his markups to a higher level at which the increased margins will offset possible sales decreases.

In any case, when overheads are not being recovered, a critical review of pricing practices is usually required and perhaps a reappraisal of the firm's objectives.

Management must always resist the temptation to allow overhead costs to build up during prosperous times. Nearly everyone enjoys spacious and luxurious offices, large staffs, new equipment, and complete inventories of materials and supplies. The contractor must realize, however, that the costs of business luxuries must be paid for out of job receipts, the same as business necessities. And when the intensity of competition increases, such luxuries add only to the firm's need for income—not to its ability to generate income. High volumes and high markups are required to support high overheads—and high markups are seldom compatible with high volumes.

Summary

Overhead costs are costs that cannot be attributed directly to specific jobs or parts of jobs. They are related instead to the contractor's capacity to do work, and may vary with the amount of work bid and the amount of work performed. There are two basic types of overhead costs: job overheads and general overheads. Job overheads are caused by specific jobs, but cannot be charged to any particular part of the job; they can be handled in the same way as direct costs. General overheads, which may be either fixed or variable, depend largely on the firm's capacity and the amount of work bid but must be spread among the jobs received. In allocating general overhead costs to the jobs received, the most common and satisfactory approaches are (1) to add a fixed percentage to total estimated direct job costs and (2) to use a percentage markup which varies according to the job's characteristics. In both cases, the appropriate percentage markup is based on the firm's anticipated sales volume, and the objective is always the same: to recover through the markup the amount of money necessary to run the business. Markups and sales volume must be carefully and continually observed to insure the recovery of all overhead costs.

11

Break-Even Analysis

*There are but two ways of paying a debt: increase of
industry in raising income, or increase of thrift in
laying out.*

THOMAS CARLYLE

Most contractors will agree that breaking even—while scarcely the ultimate objective of any businessman—is no minor accomplishment in the fiercely competitive contract construction business.

Naturally, every businessman hopes to make a profit, rather than just to break even. But identification of the firm's break-even point is nevertheless essential, for there can be no profit until this point has been reached, and failure to attain the necessary break-even volume will result in a loss.

Break-even analysis is a useful tool for the contractor in running his business and in planning for profits, for many management decisions require that the effect of changing sales volumes, markups, and overheads on the company's overall profit position be appraised. And the major problem encountered by management in making these types of decisions is in being able to visualize correctly the interactions among all the different elements that must be considered.

Break-even analysis can clearly illustrate the essential relationships among such factors as direct job costs, overhead costs, percentage markups, sales volumes, and profits. By clearly defining these relationships, management can objectively evaluate the impact of varying the different elements, and can thus select appropriate strategies for working toward the firm's objectives.

Terms Used in Break-Even Analysis

In break-even analysis, costs must be broken down into their fixed, semi variable, and variable elements.

129

Fixed costs remain constant over a period of time, regardless of the amount of work performed by the firm. Fixed costs include many of the general overhead items.

Semivariable costs may be fixed within certain ranges of sales volumes but will increase generally in some relation to sales. Semivariable costs include some of the capacity-related general overhead items, plus any additional administrative expenses necessary to handle heavy work loads. If semivariable costs vary directly with the volume of work, they can be included with variable (or direct) costs; if they are stable over well-defined ranges, they may be handled in the same manner as fixed costs.

Variable costs are those that vary directly with the amount of work undertaken. These include both direct job costs and job-related overheads.

Total income and total sales are used synonymously, referring to the total amount of money generated by the contracts received.

Total costs are made up of the sum of all fixed, semivariable, and variable costs incurred over a specified period of time.

Profit refers to the difference between total income and total cost, when income is greater than cost. Unless otherwise specified, profit is computed before income taxes.

Loss refers to the difference between total cost and total income, when cost is greater than income.

The break-even point is the sales volume at which there is neither profit nor loss, or where total income is exactly equal to total cost. The break-even point may also be expressed as a percentage of capacity rather than as a volume of sales.

The markup is the percentage added to estimated direct costs—both job costs and job overheads—to recover all other costs of running the business and to return to the owners a profit on their invested time and capital.

Information Used in Break-Even Analysis

Cost data provide the basis for break-even analysis. These basic cost data should be compiled, analyzed, and related to different levels of sales volume.

Historical accounting records usually provide the best sources of information to be used in break-even analysis, although in the absence of such records, good estimates may be equally satisfactory for planning purposes.

Table 15 shows a summary operating statement (or profit-and-loss statement) for the Aardvark Bidding Company ("Always first—in the Yellow Pages"). From these data, plus an intimate knowledge of the firm's operations, sufficient information can be obtained or inferred to perform the break-even analysis.

**TABLE 15.   Aardvark Bidding Company: Condensed
Income Statement for Year Ending December 31**

Total sales		$500,000
Cost of operations:		
Direct job costs	$407,000	
General expenses	76,000	
Total costs		483,000
Net profit before income taxes		$17,000

Basic Break-Even Arithmetic

Assuming that the "general expenses" of $76,000 shown in the operating state-
ment of Table 15 refer to the firm's fixed costs, and that the "direct job costs"
figure of $407,000 includes all variable costs, the firm's break-even point can be
quickly determined.

The break-even point, by definition, is the point at which total income
equals total cost. Total cost, in turn, includes both fixed and variable costs. To
find the break-even point for the Aardvark Bidding Company, then, the
amount of variable costs which when added to the $76,000 in fixed costs will
equal total income at that point must be calculated. In making this calculation
the total variable cost of $407,000 is assumed to have been accumulated at a
constant rate over the $500,000 sales volume; therefore, for each dollar of sales,
$407,000/$500,000, or $0.814, in variable costs are incurred. Or, stated
another way, variable costs are equal to 81.4 percent of the sales volume.

Having accumulated the necessary information, the following basic formula
can be used to calculate the break-even point:

$$V = F + D$$

where V equals the volume of sales required to break even, F is the total
amount of fixed cost, and D represents the total amount of direct costs at the
break-even point, expressed as a percentage of the sales volume.

Substituting the data for the Aardvark Bidding Company in the formula
above gives

$$V = \$76,000 + 0.814V$$
$$0.186V = \$76,000$$
$$V = \$409,000$$

The break-even point for the Aardvark Bidding Company, then, occurs at a
sales volume of $409,000. Should the company fail to achieve this volume, a

loss would result, and the firm can begin realizing a profit only after a sales volume of $409,000 has been reached. The break-even calculation can be easily checked by working backward from the break-even point as follows:

Sales volume at break-even point	$409,000
Less variable expenses at 81.4% of sales	333,000
Amount remaining to cover fixed costs	$ 76,000
Less fixed costs	76,000
Net profit or loss	$ 0

The Break-Even Chart

While the arithmetic calculations involved in break-even analysis are simple and should be thoroughly understood, the use of a break-even chart provides a more vivid picture of exactly where the firm stands with respect to its break-even point. Also, the relationships among the various factors which influence the break-even point can best be expressed graphically.

The conventional break-even chart assumes that fixed costs remain stable regardless of the level of sales, and that variable costs change in direct proportion to sales. Basically the same type of information is required to construct a break-even chart as is used in the arithmetic treatment: cost data must be accumulated and summarized for varying levels of sales.

Figure 23 shows the essential features of a break-even chart, using again the operations of the Aardvark Bidding Company (whose income statement was summarized in Table 15).

In Figure 23, the vertical axis represents both income and cost, and the horizontal axis shows total sales volume. The units of measurement on both scales are dollars.

At zero sales, income would also be zero. Therefore, the "total income" line is a straight line passing through the origin (or zero point), representing the income increasing as sales increase. Since Aardvark Bidding Company's income is derived solely from sales, the total income shown on the vertical axis will always be equal to the total sales volume shown on the horizontal axis.

Next, the total amount of fixed costs is plotted on the graph. Fixed costs in Aardvark's case total $76,000 and are assumed to remain constant at least throughout the range of sales shown on the break-even chart. The fixed-cost line, therefore, will be parallel to the horizontal axis. If some of the fixed costs are expected to increase with a larger volume of work, this fact should be reflected in the plot of fixed costs; situations involving such complications are treated later.

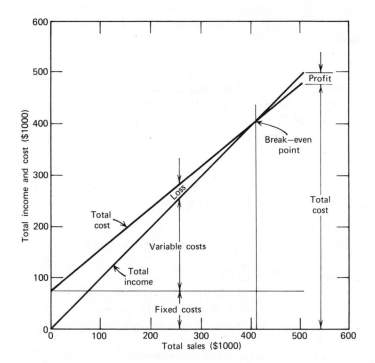

FIGURE 23. Conventional break-even chart.

Variable costs are then added to fixed costs to arrive at total costs. The slope of the variable cost line depends on the rate at which the variable costs are accrued. Variable costs are most conveniently plotted as a certain amount per dollar of sales; in this example, variable costs average $0.814 per sales dollar ($407,000 in variable costs for $500,000 in sales).

The variable costs are plotted above the fixed-cost line, thereby giving the total cost of operations foy any given level of sales. The total cost line—the sum of fixed and variable costs—indicates the total cumulative amount spent at a specific point.

The point at which the total cost line intersects the total income line—$409,000 in this example—is the break-even point. To the left of this point the vertical distance between the total income and total cost lines indicates a net loss. To the right the corresponding vertical distance between the two lines represents a net profit. Any change in sales from the break-even point, however slight, will result in either a profit or a loss.

The break-even point of the Aardvark Bidding Company occurs at a sales volume of $409,000. At this point the $409,000 in sales revenues is exactly matched by the $409,000 in costs, made up of $76,000 in fixed costs and $333,000 in variable costs. At sales volumes less than $409,000 losses will be suffered; at higher volumes, profits will be earned. Table 16 shows the effect of different sales volumes on the firm's profits.

The break-even point could be described equally well in terms of the firm's operating capacity, instead of as a dollar sales volume, although dollars are more easily defined than capacity. For example, if the firm's annual capacity were $800,000, the break-even point would have occurred at 51 percent of capacity. To earn net profits of $17,000, operations would have to be conducted at 62.5 percent of capacity. The procedure for calculating the break-even point is exactly the same as when sales volume is used; only the units of measurement along the horizontal axis of the graph would be different.

Effect of Varying Fixed Costs on the Break-Even Point

The preceding example assumed that fixed costs remained constant regardless of the firm's sales volume. While this assumption may be valid for a limited number of firms, such a condition is unlikely to exist in most contractors' operations. For as the capacity to do business increases the overhead or fixed costs will also increase.

Figure 24 illustrates the effect of varying fixed costs on the firm's break-even point. The vertical axis still represents income and cost, and the horizontal axis, total sales volume. The total income line remains the same: $1.00 of total income resulting from each dollar of sales. Also, the slope of the variable cost line is assumed to remain constant, rising $0.814 for each sales dollar.

As indicated by Figure 24, the effect of lowering the level of fixed costs is to cause the total cost line to intersect the total income line more quickly,

TABLE 16.   Aardvark Bidding Company: Net Profit or Loss Associated with Various Sales Volumes (Constant Fixed Costs)

Sales Volume	Fixed Costs	Variable Costs	Total Cost	Net Profit or (Loss)
$100,000	$76,000	$81,000	$157,000	$(57,000)
200,000	76,000	163,000	239,000	(39,000)
300,000	76,000	244,000	320,000	(20,000)
400,000	76,000	326,000	402,000	(2,000)
409,000	76,000	333,000	409,000	0
500,000	76,000	407,000	483,000	17,000

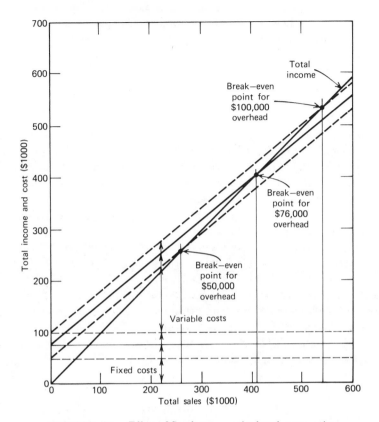

FIGURE 24. Effect of fixed costs on the break-even point.

representing a lower break-even point. In the example, lowering the fixed costs from $76,000 to $50,000 reduces the break-even point from $409,000 to $269,000. In other words, a reduction of $26,000 in fixed costs results in a reduction of $140,000 in the break-even point.

By reducing fixed costs by any given percentage, then, the break-even point can be reduced by that same percentage.

The same relationship holds true when fixed charges increase. An increase from $76,000 to $100,000 moves the break-even point farther away. In the example, the break-even point increases from $409,000 to $539,000, meaning that $130,000 in additional sales will have to be achieved to make up for the additional $24,000 in fixed cost obligations. Again, the same percentage increase must be made in sales as in fixed costs to retain the same position of economic equilibrium.

Figure 25 shows how the Aardvark Bidding Company's operation might look if all three of these levels of fixed costs—$50,000, $76,000, and $100,000—were incurred over certain ranges of sales volume.

In Figure 25, fixed costs of $50,000 are associated with sales volumes of $0 to $200,000; $76,000 in fixed costs will be incurred at sales volumes between $200,000 and $500,000; and $100,000 in fixed charges will be necessary to support sales between $500,000 and $1 million.

As is evident from Figure 25, the firm will be unable to break even under the specified conditions with fixed costs of $50,000 chargeable against the first $200,000 sales volume, for the break-even point would not be reached until a sales volume of $269,000 had been achieved—and any volume above $200,000 requires taking on additional fixed costs.

Over the second range of sales volumes—$200,000 to $500,000—the break-even point will occur at $409,000 with $76,000 in fixed costs. The net profit will rise to $17,000 at $500,000 in sales, made up of the difference between the $500,000 income and the $76,000 in fixed costs and $407,000 in variable costs.

At any point over $500,000 volume, however, fixed costs increase to $100,000 and the break-even point rises to $539,000. To break even, then, requires $130,000 more volume than before, and to make the same amount of profit as was made at $500,000 sales volume ($17,000) would now require sales of $630,000, also $130,000 more than before. Table 17 shows the firm's profit position at various sales volumes.

Effect of Markups on the Break-Even Point

The markup—or percentage added to direct job costs to cover overhead and profit—has an important effect on the volume of work required to break even. The higher the markup, the lower the break-even point; and the lower the markup, the higher the break-even point.

Figure 26 shows how the break-even point varies with the markup. In Figure 26, the firm has annual fixed costs of $50,000. With a 10 percent markup the break-even point occurs at a sales volume of $550,000. This volume is matched by total costs of $550,000, made up of $500,000 in direct (or variable) costs plus the 10 percent markup which covers the $50,000 in fixed costs. The variable cost line has a slope of $500,000 in $550,000, or $0.909 per dollar of sales.

A higher markup decreases the slope of the variable cost line, causing it to intersect the total cost line more quickly. A 20 percent markup results in a break-even point of $300,000, representing a slope of $250,000 in $300,000, or $0.833 per sales dollar.

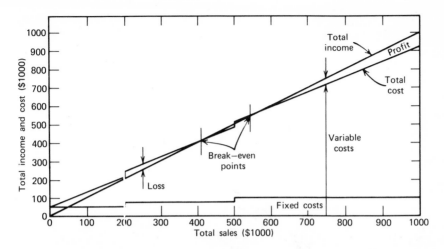

FIGURE 25. Break-even chart with varying levels of fixed costs.

Similarly, a lower markup, by raising the slope of the variable cost line, causes its intersection with the total income line to come later. For a 5 percent markup the break-even point is at $1,050,000, made up of $1,000,000 in direct costs and $50,000 in fixed costs. Variable costs in this case are incurred at a rate of $0.952 per dollar of sales.

TABLE 17. Aardvark Bidding Company: Net Profit or Loss Associated with Various Sales Volumes (Varying Levels of Fixed Costs)

Sales Volume	Fixed Costs	Variable Costs	Total Cost	Net Profit or (Loss)
$100,000	$50,000	$81,000	$131,000	$(31,000)
200,000	50,000	163,000	213,000	(13,000)
300,000	76,000	244,000	320,000	(20,000)
400,000	76,000	326,000	402,000	(2,000)
409,000	76,000	333,000	409,000	0
500,000	76,000	407,000	483,000	17,000
539,000	100,000	439,000	539,000	0
600,000	100,000	488,000	588,000	12,000
630,000	100,000	513,000	613,000	17,000
700,000	100,000	570,000	670,000	30,000

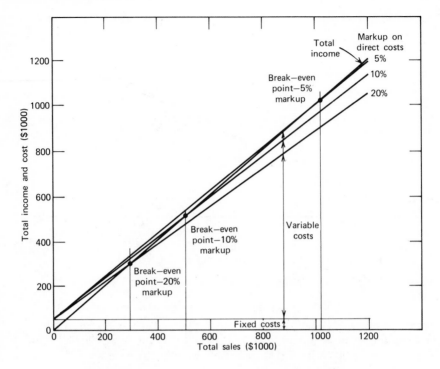

FIGURE 26. Break-even chart with varying markups.

Derivation of the Break-Even Formula

The mathematical relationships among fixed charges, percentage markups on direct job costs, and the volume of work required to break even, can be easily derived

By definition the break-even volume (V) occurs at the point at which the sales volume, or total income, equals the total cost. The total cost consists of the sum of direct or variable costs (D), and indirect or fixed costs (F). Thus

$$V = D + F$$

Since fixed charges are recovered by applying some percentage markup (M) to the direct costs, the following relationship must exist at the break-even point:

$$F = \frac{M \times D}{100} \quad \text{or} \quad D = \frac{100F}{M}$$

Substituting for D in the first equation, the break-even volume is found to occur at

$$V = \frac{100F}{M} + F = F\left(\frac{100}{M} + 1\right)$$

The ratio of the break-even volume to the fixed charges is therefore

$$\frac{V}{F} = \frac{100}{M} + 1$$

where M is expressed as a percentage.

For example, using this formula, a 10 percent markup will require a sales-fixed charges ratio of $(100/10) + 1$, or 11 times. Similarly, a 20 percent markup will require a total sales volume of $(100/20) + 1$, or 6 times the fixed charges. Table 18 shows the sales dollars required per dollar of overhead for various markups while the same relationships are pictured graphically in Figure 27.

Relationships Among Factors Affecting the Break-Even Point

The break-even point depends both on the contractor's fixed costs and percentage markups on direct job costs, and any change in either fixed costs or

TABLE 18.   Sales Required per Dollar of Overhead at Various Markups

Percentage Markup on Direct Cost	Sales Required Per Dollar of Overhead	Percentage Markup on Direct Cost	Sales Required Per Dollar of Overhead
1	$101.00	12	$9.33
2	51.00	15	7.67
3	34.33	20	6.00
4	26.00	25	5.00
5	21.00	30	4.33
6	17.67	35	3.86
7	15.29	40	3.50
8	13.50	45	3.22
9	12.11	50	3.00
10	11.00	100	2.00

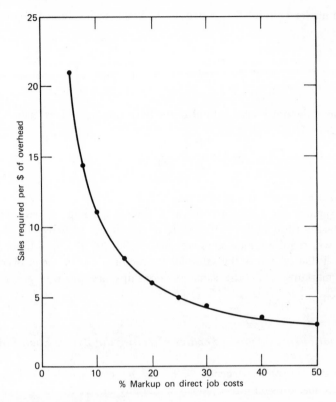

FIGURE 27. Sales required per dollar of overhead at various markups.

markups will be greatly amplified in the volume of work required to break even. Working on a 10 percent markup, each dollar of overhead must be matched by $11.00 of sales. This relationship means that if an additional $1000 of overhead expense is taken on, an additional sales volume of $11,000 must be earned, just to stay in the same position with respect to profits.

On the other hand, if overheads can be reduced by $1000 without changing the markup, the breakeven volume will be $11,000 less. If, at the same time, the sales volume could be maintained, the $1000 reduction in overheads would result in an additional $1000 in profits.

Similar effects could be brought about by varying the percentage markup while keeping overheads and sales constant, or by varying the sales volume with a fixed markup and overhead. One contractor may be able to reduce his overhead without difficulty, while another will find it easier to raise his volume slightly, or to increase his markup by 1 or 2 percent.

In any event, only a small change in the contractor's markup, overhead cost, or sales volume makes a tremendous difference both in the volume of work he must obtain to break even and in his ultimate profit position.

The relationships among these factors explain to some extent how certain contractors are able to run more profitable businesses on $100,000 sales volumes than others are able to manage with million-dollar volumes. A good markup on a small volume is apt to be far more profitable, and far less risky, than a large volume accumulated by means of low markups; for no amount of volume can compensate for the lack of a profit. The small-volume contractor with a large profit is either working on a high markup or a low overhead; the end result is the same in either case.

Summary

A firm's break-even point is the volume of sales at which there is neither profit nor loss; break-even analysis refers to the procedures used in identifying this point. Break-even analysis also provides a wealth of useful information regarding the relationships among a contractor's fixed costs, variable costs, percentage markups, and profits. These relationships are most conveniently expressed in graphic form, by means of a break-even chart. Fixed costs and percentage markups on direct job costs exert the strongest influence on a firm's break-even point. The sales volume required per dollar of fixed cost can be calculated from the following formula:

$$\frac{V}{F} = \frac{100}{M} + 1$$

where V is the volume of work required to break even; F is the amount of fixed costs; and M is the percentage markup on direct costs. As is evident from this formula, a reduction in the markup greatly increases the amount of work required to break even; and any increase in fixed costs should be accompanied by proportionate increases in sales.

12

Profits and Profit Analysis

Profit is the product of labor plus capital multiplied by management. You can hire the first two. The last must be inspired.

FOST

The economist's definition of business is "commercial or financial affairs conducted for the sake of profit." Business activities are thus distinguished from other types of economic activities which may be concerned with making a living, but not necessarily a profit.

Profits are not just incidental to a business; they are what makes the business a business. And while profits in some instances may not be the only, or even the primary, objective of a firm's operations, profits are nevertheless essential for the firm's economic survival and for its pursuit of any other, though less tangible, objectives.

Profit is an attitude. Although most segments of the construction industry have been plagued by profit problems for some time, a company's profits, whether in construction or any other field, are still more dependent on the company's management than upon the industry in which it operates. Every industry has its share of highly profitable operations. The fact that the construction industry has a relatively large proportion of marginal ones is at least as much a reflection on the attitude of construction management as on the industry itself. There is not such thing as an unprofitable industry; there are only unprofitable companies operating within industries.

Probably the main cause of unprofitable operations is management's refusal to be profit-minded. The tendency, instead, is to operate on hope, uncontained optimism, and intuition—rather than on facts about the way things really are.

Only by objectively viewing profits and profit potentials in light of the company's own capabilities and limitations and considering these factors in respect to the industry as a whole can management hope to realize the company's optimum profits.

And profits are the ultimate measure of management effectiveness.

The Nature of Profits

Profits are easy enough to define, even though their adequacy is hard to measure. Profits, simply stated, represent the net difference between total income and total cost. Profits are the result of a firm's operations and a measure of the economic success of its operation.

Profits are influenced by a number of different factors, both internal and external. Internal factors refer to the company's operations; external factors are brought about by the industry in which the company operates and by the nature of the competition that is encountered.

The contracting industry probably comes closer to having "perfect competition" than any other modern industry—competition in which each seller has little or no control over prices. In such a profit-and-loss system (*not* just a profit system), the contractor is forced to get the most from his available resources.

The primary requirements for profitable operations in the contracting business are to first select jobs on the basis of their profit potential, and then to conduct operations with the greatest possible economy of performance.

Profits should be looked on as a kind of cost. At a bare minimum, profits must cover the cost of supplying capital for the business. Additionally, profits should cover the cost of personally contributed resources, offer a reward for entrepreneurship, reimburse the risk-taker for the chances he takes, and, in some cases, offer extra returns for some types of monopoly position.

Personally Contributed Resources

Personally contributed resources include the capital, time, and effort of the owners of the business. Profits from operations must be sufficient to cover the costs of these resources, or to equal the income that could be earned from them elsewhere, or the business cannot be justified on an economic basis. Interest must be earned on capital; if the profits from the business are not at least equal to the interest that could be earned from the capital in comparable alternative investments, then the capital has obviously been been misinvested. Similarly, skilled management is entitled to high wages, since good management is scarce and in great demand; if the business does not reimburse management with as high a level of wages as could be earned elsewhere, the management is misplaced.

Reward for Entrepreneurship

The entrepreneur is defined as "an employer of productive labor; a contractor; one who undertakes to carry out any enterprise." The entrepreneur—the man who through his imagination, skill, daring, and fortitude is willing to take the inherent risk of being on his own—is entitled to rewards above what he would earn as a skilled manager working for others.

Compensation for Risk and Uncertainty

Profits are linked closely with the elements of risk and uncertainty. If there were no risk associated with construction work, neither would there be much opportunity for profit; for the fear of risking their resources is what discourages many people—even those having the necessary financial and technical competence—from engaging in the contracting business. Always, the profits on good jobs must more than offset the losses on bad ones; being able to do this entails considerable judgment and experience on the part of management in anticipating the unexpected and unpredictable.

Bonus for a Monopoly Position

Some fortunate firms may find themselves in a monopoly position in that they are the only ones capable of performing a certain type of work, or they may be the only ones available to do it, or they may have a reputation for being able to do it better than anyone else. Being in a monopoly position enables the contractor to price his work for "what the traffic will bear." His competition will not be with other firms so much as with whether the project will be undertaken at all. The true monopoly position should be distinguished, however, from the job that no one else is willing to undertake because of excessive risks.

Measuring Profits and Profitability

A distinction should be made between profits and profitability. Profits refer to the difference between total income and total cost and are therefore expressed in terms of dollars. Net profit, as used throughout this book, refers to the profit before income taxes but after all other deductions.

Profitability is a measure of the desirability of risking additional capital and expense in undertaking new projects, and it is also a measure of the efficiency with which the firm's resources have beem employed. Return on investment is the most important measure of profitability.

Again, profits *always* refer to the difference between *total* income and *total* cost. There is a tendency on the part of some companies (and on some labor unions) to define profits in terms of "cash flow"—after-tax profits plus depreciation and amortization allowances. Cash flow is undeniably an important measure of a company's ability to invest in new equipment and to expand its operation, since cash flow represents the actual amount of money generated by the business and available to it. But depreciation is equally undeniably a cost, and costs are the direct enemies of profits. *Cash flow is not profit.*

Probably the only valid generalization regarding a firm's profits is that profits are not as high as they should or could be. But the chances for raising profits to their optimum level are better if they are planned and measured as a percentage of the capital employed in the business.

What constitutes a satisfactory percentage return on invested capital has never been, and will probably never be, answered, for the adequacy of profits is not a measurable quantity and can be judged as satisfactory or unsatisfactory only by comparison with others. Some of the major considerations in determining the adequacy of profits are the profit levels of all other companies, the profit levels within the industry and similar industries, and the company's position and experience within its own industry.

According to Dun & Bradstreet, the minimum level of profitability sufficient to insure continued company growth is 10 percent of the firm's net worth. But most contractors should be able to double this return, since by the very nature of the contracting business the net worth is generally low in relation to the volume of business conducted.

Profit Analysis

The techniques of profit analysis are similar to the techniques of break-even analysis. For break-even analysis is, in effect, a special case of profit analysis in which the objective is to identify the point at which profits are zero. In profit analysis, relationships between profits and sales volumes are sought. As is the case with break-even analysis, the most meaningful and useful information can be obtained in profit analysis by presenting the data in graphical form, where relationships between different variables can be easily visualized.

Summarizing again last year's income statement for the Aardvark Bidding Company (from Table 15), we obtain the following:

Total sales		$500,000
Cost of operations:		
Direct job costs	$407,000	
General expense	76,000	
Total costs		483,000
Net profit before income tax		$17,000

These data can be treated in generally the same way as in the break-even analysis of Aardvark's operations in the preceding chapter. Figure 28 shows a profit/volume chart for the Aardvark Bidding Company.

In the profit/volume chart, the vertical axis is divided into two parts, the top part representing a net profit and the lower segment representing a net loss.

The horizontal axis is drawn through the zero profit point on the vertical axis, representing the break-even point. Units along the horizontal axis measure the company's total sales volume.

The profit/volume chart (or the P/V chart), as its name implies, indicates the net profit or net loss associated with any given level of sales.

At zero sales, the net loss will be equal to the amount of fixed expense. The

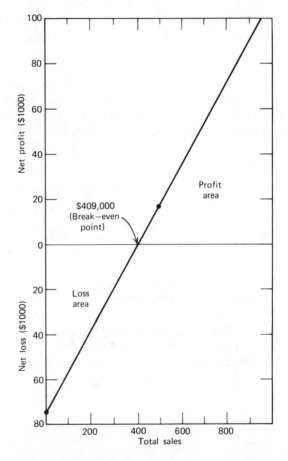

FIGURE 28.   Profit/volume chart for Aardvark Bidding Company.

P/V line, then, intersects the vertical axis at a net loss of $76,000, corresponding to zero sales volume.

Aardvark's net profit at a sales volume of $500,000 was $17,000; this $17,000 can be plotted above the $500,000 point on the horizontal axis.

A straight line connecting these two points will, presumably, give the amount of net profit or net loss corresponding to any chosen level of sales, assuming that all other factors—such as the level of markups, the amount of variable cost per dollar of sales, and the amount of fixed costs—remain the same. The point at which the P/V line intersects the horizontal axis—at $409,000 in this case—is the break-even point. The area encompassed by the P/V line and the two axes to the left of the break-even point represents net losses; and the area bounded by the P/V line and the axes to the right of the break-even point represents the firm's profits.

Profits, then, can be easily projected for any desired or anticipated sales volume, again assuming that the relationships between all the different factors stay the same. For the Aardvark Bidding Company, sales of $500,000 brought in net profits of $17,000. A sales volume of $600,000 should therefore yield $35,000 in profits; further increases in sales to $700,000 will result in $19,000 additional profits, totaling $54,000.

As is apparent from the P/V chart, an increase of any given amount in sales will bring about a uniform increase in net profits. For Aardvark, each $100,000 change in sales volume results in approximately $19,000 change in profits.

Application of the P/V Ratio

The P/V ratio provides a convenient means of measuring the firm's profit margin—the difference between sales and variable expenses.

In the preceding example, each $100,000 in sales was accompanied by a marginal income of $18,600. The P/V ratio, equal to incremental (or marginal, or additional) income divided by incremental sales, is therefore 18.6 percent.

Having established the P/V ratio, several additional calculations can be made from it.

The Break-Even Point

The break-even point can be found by dividing total fixed costs by the P/V ratio. For example,

$$\text{break-even point} = \frac{\$76,000}{0.186} = \$408,600$$

Fixed Expenses

Fixed expenses, expressed as a percentage of sales, can be found for any given sales volume by subtracting the P/V ratio from the net profit/net sales ratio:

$$\text{fixed expenses} = \frac{\$17,000}{\$500,000} - 0.186 = 0.340 - 0.186$$
$$= 0.154 = 15.4\% \text{ of sales of } \$500,000.$$

Variable Expenses

Variable expenses, as a percentage of sales, are equal to 100 percent minus the P/V ratio:

$$\text{variable expenses} = 100.00 - 18.6 = 81.4\% \text{ of sales}$$

Net Profit

The profit at any given sales volume can be found by multiplying the sales by the P/V ratio, and deducting fixed expenses:

$$\text{net profit at } \$500,000 \text{ sales} = (\$500,000 \times 0.186) - \$76,000$$
$$= \$93,000 - \$76,000 = \$17,000.$$

Margin of Safety

The margin of safety refers to the difference between the firm's break-even point and its sales volume. As such, the margin of safety reflects the volume decline that can be withstood by the company before losses are incurred. The margin of safety is found simply by taking the anticipated sales volume and subtracting from it the break-even volume, calculated as described above.

Factors Affecting Profits

The preceding analyses are valid as long as the relationships between fixed costs, variable costs, and markups remain constant. But when the contractor's percentage markups on direct job costs change, his entire profit situation also changes.

When the level of fixed costs and markups are known or can be estimated with reasonable accuracy, the amount of profit associated with any given sales volume can be quickly and easily determined.

The total sales volume (S) represents simply the sum of direct costs (D) and the markup (M) added to these direct costs. Or, stated algebraically,

$$S = D + DM = D(1 + M)$$

where M is expressed as a decimal rather than as a percentage. Solving for D in this equation gives

$$D = \frac{S}{(1 + M)}$$

Since total costs (C) consist of the sum of fixed costs (F) and direct costs (D),

$$C = F + D = F + \frac{S}{(1 + M)}$$

Profits (P), the difference between total sales (S) and total costs (C), are therefore

$$P = S - C = S - \left[F + \frac{S}{(1 + M)} \right] = S \left(\frac{M}{(1 + M)} \right) - F$$

This formula is extremely useful to the contractor in identifying the effect of changes in the average percentage markup on his firm's potential at any given sales volume. The factor $M/(1 + M)$ is referred to as the "profit factor." Table 19 gives the profit factors for selected markups ranging between 1

TABLE 19.   Profit Factors for Various Markups

Percentage Markup on Direct Cost	Profit Factor	Percentage Markup on Direct Cost	Profit Factor
1	0.0099	12	0.1071
2	0.0196	15	0.1304
3	0.0291	20	0.1667
4	0.0385	25	0.2000
5	0.0476	30	0.2308
6	0.0566	35	0.2593
7	0.0654	40	0.2857
8	0.0741	45	0.3103
9	0.0826	50	0.3333
10	0.0909	100	0.5000

percent and 100 percent (or between 1.01 and 1.00 expressed as a decimal). This profit factor, multiplied by the company's anticipated sales volume, and less overhead or fixed costs, gives the net profit to be achieved at any combination of sales volume and markup.

To illustrate the application and importance of this formula, we refer again to the operations of the Aardvark Bidding Company for whom annual fixed costs of $76,000 are expected, this time assuming an average markup of 10 percent of estimated direct job costs can be achieved. Then, if a total sales volume of $1,000,000 were anticipated, the profit could be calculated as follows:

$$\text{net profit} = (\$1,000,000)\,(0.0909) - \$76,000 = \$90,900 - \$76,000 = \$14,900$$

If the sales volume, instead of totaling $1,000,000 came to only $700,000, the net profit would amount to

$$\text{net profit} = (\$700,000)\,(0.0909) - \$76,000 = 63,600 - \$76,000 = -\$12,400$$

In this case, then, a markup of 10 percent associated with a sales volume of $700,000 would result in a net loss.

Increasing the markup to 20 percent (with a corresponding profit factor of 0.1667) would alter the situation considerably:

$$\text{net profit} = (\$700,000)\,(0.1667) - \$76,000 = 116,700 - \$76,000 = \$40,700$$

And achieving a sales volume of $456,000, still with a 20 percent markup, would result in

$$\text{net profit} = (456,000)\,(0.1667) - \$76,000 = \$76,000 - \$76,000 = 0$$

Therefore, $456,000 in sales represents the break-even point for a 20 percent markup and fixed charges of $76,000—the break-even point occurring at 6 times the overhead.

Changing the markup also changes the profit picture significantly. Table 20 shows the net profit realized on various sales volumes between $100,000 and $2,000,000 for markups of 5, 10, 15, 20, and 25 percent, using Aardvark's fixed costs of $76,000.

As indicated in Table 20, the same $1,000,000 sales volume that results in a net loss of $28,400 at a 5 percent markup, will bring in net profits of $124,000 with a 25 percent markup. Figure 29 shows graphically the profit/volume relationships for various markups. From Figure 29, it is evident that a sales volume

TABLE 20.   Aardvark Bidding Company: Net Profit (or Loss) Associated with Various Sales Volumes and Markups

Sales Volume ($)	Fixed Costs ($)	Percentage Markup on Direct Cost				
		5.0 ($)	10.0 ($)	15.0 ($)	20.0 ($)	25.0 ($)
0	76,000	(76,000)	(76,000)	(76,000)	(76,000)	(76,000)
100,000	76,000	(71,200)	(66,900)	(63,000)	(59,300)	(56,000)
200,000	76,000	(66,500)	(57,800)	(49,900)	(42,700)	(36,000)
300,000	76,000	(61,700)	(48,700)	(36,900)	(26,000)	(16,000)
400,000	76,000	(57,000)	(39,600)	(23,800)	(9,300)	4,000
500,000	76,000	(52,200)	(30,600)	(10,800)	7,400	24,000
600,000	76,000	(47,400)	(21,500)	2,200	24,000	44,000
700,000	76,000	(42,700)	(12,400)	15,300	40,700	64,000
800,000	76,000	(39,900)	(3,300)	28,300	57,300	84,000
900,000	76,000	(33,200)	5,800	41,400	74,000	104,000
1,000,000	76,000	(28,400)	14,900	54,400	90,700	124,000
1,200,000	76,000	(18,900)	33,100	80,500	124,000	164,000
1,400,000	76,000	(9,400)	51,300	106,600	157,300	204,000
1,600,000	76,000	200	69,400	132,600	190,700	244,000
1,800,000	76,000	9,700	87,600	158,700	224,000	284,000
2,000,000	76,000	19,200	105,800	184,800	257,300	324,000

that is highly profitable at one markup may be highly unprofitable at a lower markup.

In Figure 30, the effect of the markup on the profit/volume relationship is presented in a slightly different way, giving the net profit resulting from a fixed volume at any selected markup. This figure illustrates, for example, that an $800,000 sales volume acquired at a 25 percent markup will yield approximately the same net profit as a $1,200,000 sales volume taken at a 15 percent markup, or as a $2,000,000 volume at about a 9 percent markup.

The ratio of net profit to net sales can also be found from the profit formula (again with the markup expressed as a decimal):

$$P = S\left(\frac{M}{1 + M}\right) - F$$

Dividing both sides of the equation by S gives

$$\frac{P}{S} = \left(\frac{M}{1 + M}\right) - \frac{F}{S}$$

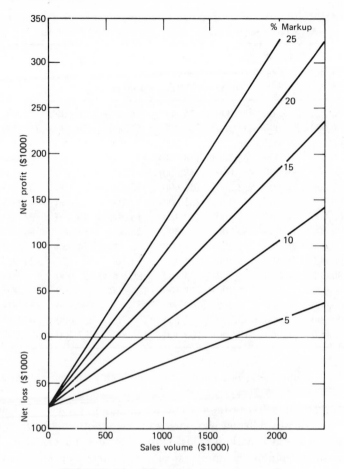

FIGURE 29.   Effect of markup on net profits.

At a sales volume of $1,000,000 and a markup of 10 percent (0.1), Aardvark Bidding Company will realize a net profit on sales of

$$\frac{P}{S} = \frac{1.0}{11.0} - \frac{\$76,000}{\$1,000,000} = 0.0909 - 0.0760 = 0.0149 = 1.49\%$$

Net profits as a percentage of net sales are shown graphically in Figure 31, as a function of the sales/fixed cost ratio and the percentage markup. A firm using an average markup of 15 percent on its direct job costs, in order to realize a margin of 4 percent on its sales, must have sales totaling at least 11

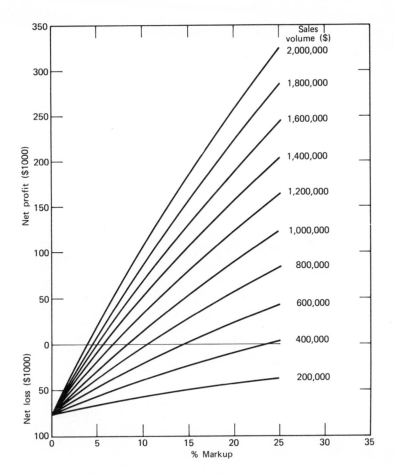

FIGURE 30. Effect of sales volume on net profits.

times its overhead. Similarly, a return of 2 percent on sales would require a total volume of about 9 times the firm's fixed costs at the same 15 percent markup.

The Laws of Profits

Some general observations and conclusions can be drawn from the preceding analyses of relationships among profits, sales costs, variable costs, and markups. These findings can be summarized as follows, representing the "laws

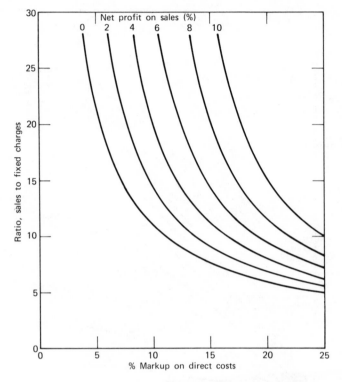

FIGURE 31. Relationships between sales, fixed costs, markups, and profit margins.

of profits" for contractors:

- The first law of profits is that *all* costs—not just out-of-pocket costs—must be included when determining the firm's profits.
- Any change in fixed costs will change the net profit by a like amount if all other cost relationships remain constant. Each dollar increase in fixed costs results in a dollar's decrease in profits; each dollar decrease in fixed costs increases profits by a dollar.
- Any percentage change in fixed costs will change the firm's break-even point by the same percentage, and in the same direction, assuming other cost relationships hold steady.
- A change in the rate at which variable costs are incurred will change both the break-even point and the net profit earned at any specified sales volume, for a constant level of fixed costs. An increase in the variable

expense rate will raise the break-even point and lower profits; decreasing the rate will lower the break-even point and increase profits.

- An increase in the percentage markup on direct job costs will lower the break-even point and increase the net profit earned at any volume. A decrease in the markup will raise the break-even point and lower the profit realization at any sales volume.

- A change in both fixed costs and variable costs will have a significant effect on profits if they both change in the same direction, but a less noticeable effect if they move in opposite directions.

Summary

Profits are essential in every business. An adequate level of profits must return to the contractor-owner the cost of his personally contributed resources, a reward for his entrepreneurship, compensation for the risks he takes, and in some cases an additional bonus for his being in a monopoly position. Profits represent the difference between total income and total cost, measured in dollars; profitability measures the efficiency with which the company's resources have been used and is generally measured as a percentage return on investment. Profit analysis employs many of the tools of break-even analysis, the objective usually being to relate net profits to the company's anticipated sales volume. Relationships between the many different factors affecting profits are usually expressed graphically by means of a profit/volume chart, which can be constructed to show the impact of changing markups, sales volumes, and profits. The profit/volume ratio can be used in different ways to find the firm's break-even point, variable expense rates, fixed expenses, or profits, given a uniform profit/volume relationship. Changes in markups will affect the profits significantly, and the impact of prices on profits should be examined critically. The laws of profits state, in general terms, the cause-and-effect relationships among fixed costs, variable costs, markups, and profits.

13

How to Figure the Odds

It is a truth very certain that, when it is not in our power to determine what is true, we ought to follow what is most probable.

DESCARTES

Almost all management decisions are based on beliefs—or on judgments of probabilities—rather than on certainties, with experience usually providing the basis for these judgments. Certainties are scarce enough in most businesses, and are almost nonexistent in the contracting field. Fortunately, the application of statistical techniques and probability theory can do much to help define uncertain events in objective, measurable terms.

Nevertheless, many contractors are inclined to show a complete disregard for, and even a distrust of, statistical methods; they believe apparently that the traditional ways of doing things are the best ways, and that any application of scientific or analytical techniques reflects on their ability as managers to make intuitive decisions. What they are saying in effect is that their operations are so crude and undeveloped that they cannot benefit from the powerful new management tools that have helped so many other industries and businesses.

This is not true, for the contracting business like every other business is concerned with the management of manpower, machines, materials, and money. These resources can be handled effectively, or they can be handled poorly. But if any problem involving any of these resources can be solved better by using scientific techniques than by making "hunch" decisions, then the only sensible and intelligent way to solve the problem is by using the best techniques available.

A company's success depends primarily on the knowledge and judgment of its management. And a sound understanding of statistics and statistical techniques can provide the necessary knowledge to sharpen management's judgment by removing much of the necessity for guesswork and speculation and thereby substituting a solid foundation for making sound management decisions.

The Use of Statistics

Statistics is defined as "the science which deals with the collection, classification, and use of numerical facts or data bearing on a subject or matter."

Statistics, therefore, consist of organized facts, and the objective of statistical analysis is to select and analyze facts to arrive at useful conclusions. If the right type of data have been selected, and if these data have been analyzed properly, then the conclusions thus derived can be applied to the future as well as the past in making predictions of things to come.

The usefulness of statistics as a tool for construction management will depend to a large extent on the availability of suitable data, for the inferences to be drawn from statistical analysis can be no better than the raw data used as a basis for the study. Numerical data suitable for statistical treatment include information on sales, estimated costs, actual costs, profits, competitors' prices, and so on—almost anything that can be measured in dollars.

Whenever large quantities of numerical facts need to be dealt with, statistical techniques should be employed. To attempt to draw meaningful conclusions and formulate company policies on the basis of past experience without using sound statistical methods is committing the company to work under an unnecessary, and often serious, handicap. Probably the only reason that more contractors have not suffered more from their failure to employ statistical methods as a management aid is that most of their competitors have also failed to avail themselves of this opportunity.

In accumulating data for analysis and for eventual use in business planning, the data must be arranged and grouped in some logical, easy-to-interpret fashion. Usually this requires the compilation of frequency distributions and the construction of a frequency distribution curve.

Frequency Distribution

A frequency distribution is simply a grouping of statistical data after the data have been sorted and arranged in some logical order, such as from the smallest to the largest. The purpose of the frequency distribution is to show the frequency with which the data in each group appear.

The frequency distribution is particularly valuable to the contractor in analyzing such relationships as those between his own estimated and actual costs on work he has performed, in comparing his own estimated costs with his competitors' bids on different types of jobs, and in making other kinds of comparisons involving large quantities of data that might otherwise be difficult to interpret.

Often, data are such that simple arithmetic averages or any other single statistical measure is either meaningless or of only limited value. With the frequency distribution, in addition to identifying the average values of a mass of data, the entire range of values can be clearly defined.

In constructing a frequency distribution curve, raw data are grouped into their appropriate classifications by dividing the overall range of values covered by the data into convenient-size groups, and then by tallying the data falling within each group. Table 21 illustrates some grouped data used in analyzing the bids of a large number of competitors accumulated over many jobs.

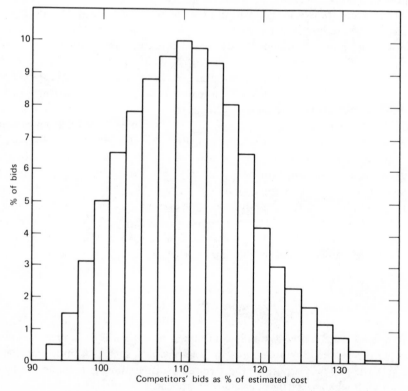

FIGURE 32.   Frequency distribution of competitors' bids.

In Table 21, 1000 competitors' bids have been converted to a percentage of the estimated direct job costs for the jobs on which the bids were submitted. These percentages were then grouped into narrow ranges—91 to 93 percent, 93 to 95 percent, and so on up to 135 to 137 percent, thereby covering the entire range of bids. Of the 1000 bids that were included in the analysis, 100 fell between 109 and 111 percent of the estimated direct job costs, 50 were between 99 and 101 percent of estimated cost, 5 were between 93 and 95 percent, and 12 were in the 127 to 129 percent range.

Figure 32 shows the frequency distribution in graphical form as it is plotted from the data in Table 21. This figure is generally typical of frequency distribution curves in that beginning from zero the curve rises to a maximum value somewhere around its average value, then tapers off and finally reaches zero again. If the curve is perfectly symmetrical, it is usually referred to as a normal curve or a normal distribution, and the highest point on the curve will correspond to the arithmetic average of all the data. In statistical terms, the distribution of Figure 10 is "positively skewed," signifying that the distribution tails off toward the right at a slower rate than toward the left, and meaning that the range of values on the high side of the average value is greater than the range of values on the low side.

The Cumulative Frequency Distribution

The right-hand column in Table 21 represents the total number of bids equal to or greater than the range given in the left-hand column. This total is found by adding together the total number of bids falling within each range, starting from the end and working backward. Since there was one bid falling in the 133 to 135 percent range and four in the 131 to 133 percent range, the total number of bids equal to or above 131 percent of estimated direct cost was five. Similarly, a total of 13 bids fell above 129 percent of estimated cost, and altogether 25 bids were more than 127 percent of cost. All 1000 of the bids were above 93 percent of cost.

The cumulative total number of bids falling above each range of value is plotted graphically in Figure 33. As is apparent from the figure, the frequency distribution curve when plotted cumulatively starts at 1000, or at 100 percent of the total number of bids included in the analysis. From this point the curve decreases rapidly until it reaches approximately the middle, at which time the rate of decline decreases and the curve tapers off gradually until reaching zero. Actually, the exact point at which the rate of decline changes corresponds to the maximum value in the frequency distribution from which the cumulative distribution was plotted—in this case, the 109 to 111 percent range of Figure 32.

The cumulative distribution curve will be especially valuable in analyzing the bidding characteristics of specific competitors (this is covered in subsequent chapters). The cumulative frequency distribution curve can be used in predicting the probability of an event's occurrence, assuming that historical data are indicative of the future—or, more specifically, assuming that a competitor will behave in the future in approximately the same way he has behaved in the past. For example, experience has shown this contractor that approximately 5 percent of the bids submitted by his competitors were below estimated cost; therefore, even by bidding jobs at cost, this contractor could expect to underbid only 95 percent of his competitors.

TABLE 21.   Frequency Distribution Data

Competitors' Bids as Percentage of Estimated Job Cost	Total Number of Bids	Cumulative Total Number of Bids Equal or Above
91 to 93	0	1000
93 to 95	5	1000
95 to 97	15	995
97 to 99	31	980
99 to 101	50	949
101 to 103	65	899
103 to 105	78	834
105 to 107	88	756
107 to 109	95	668
109 to 111	100	573
111 to 113	98	473
113 to 115	93	375
115 to 117	80	282
117 to 119	65	202
119 to 121	42	137
121 to 123	30	95
123 to 125	23	65
125 to 127	17	42
127 to 129	12	25
129 to 131	8	13
131 to 133	4	5
133 to 135	1	1
135 to 137	0	0
Total	1000	

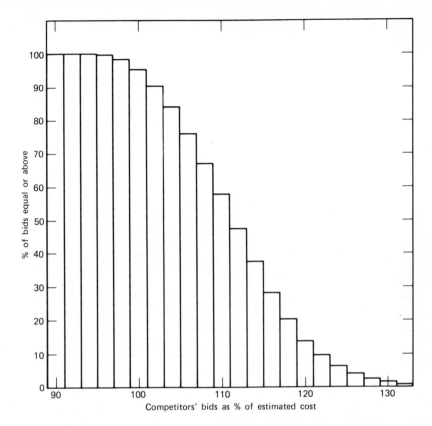

FIGURE 33.   Cumulative frequency distribution of competitors' bids.

The Theory of Probability

Probability is a fascinating subject, a fact to which any card or dice player will attest. Probability theory also plays an increasingly important role in business decisions.

Most management decisions require that the probable outcomes of different events be predicted. And since predictions of future events can never be made with complete certainty, the most successful managers are those who can make correct predictions more often than their competitors.

Probability refers to the frequency with which an event can be expected to occur, or to the likelihood that a particular event will happen. Probability is measured as the ratio of the number of different ways that a particular event can happen to the total number of possible outcomes.

For example, there is but one way that a single, six-sided die can come up 6, while there are a total of six different possible outcomes for a single throw of the die; the probability of throwing a 6 is therefore ⅙.

Similarly, there are 36 different combinations possible with two dice:

1–1	2–1	3–1	4–1	5–1	6–1
1–2	2–2	3–2	4–2	5–2	6–2
1–3	2–3	3–3	4–3	5–3	6–3
1–4	2–4	3–4	4–4	5–4	6–4
1–5	2–5	3–5	4–5	5–5	6–5
1–6	2–6	3–6	4–6	5–6	6–6

Of these 36 equally likely possibilities, six different combinations will result in a total of 7: 1–6, 2–5, 3–4, 4–3, 5–2, and 6–1. The probability of throwing a 7, then, is ⁶⁄₃₆, or ⅙. There is only one combination of two dice that will result in either a 2 or 12; two different ways of making a 3 or 11; three ways of throwing 4 or 10; four combinations that will result in 5 or 9; and five that will give 6 or 8.

Since there are six ways of throwing a 7 to only four ways of throwing a 5, the probability that a 7 will appear before a 5 does is $6/(6 + 4)$, or .6. If the game is craps and the point is 5, the odds are 6 to 4 (or 3 to 2) against the player: There are six different ways he can lose by throwing a 7, against only four possible ways to win by throwing a 5.

Probabilities are expressed as a ratio, or fraction, between 0.0 and 1.0, or as a corresponding percentage from 0 to 100. A probability of 0 denotes an event that cannot happen, while a probability of 1.0, or 100 percent represents absolute certainty. Everything else falls between these two extremes.

The "theory of probability" is really nothing more than common sense expressed in numbers. By defining the likelihood of an event's occurrence in numerical terms, a relatively precise measure is possible of many situations that could otherwise be measured only instinctively or intuitively.

The Rules of Probability

The most important principles of probability can be summarized in six general rules. These six rules of probability are adequate for solving almost any problem involving probability, whether the problem involves gambling or business.

Rule 1

Probabilities are always between 0.0 and 1.0. This first rule is essentially a part of the definition of probability.

Rule 2

The probability that an event will happen, plus the probability that it will not happen, always equals 1. This rule follows logically from the first.

Rule 3

The probability that one or the other of two mutually exclusive events will happen is equal to the sum of their separate probabilities. "Mutually exclusive" means simply that both events cannot happen at the same time; since having both a 1 and a 2 show up on one throw of a single die is not possible, these two events are mutually exclusive. And since the probability of either a 1 or a 2 appearing individually is ⅙, the probability of one or the other coming up in a single throw is ⅙ + ⅙, or ⅓.

Rule 4

The probability that two or more events will happen simultaneously is equal to their individual probabilities all multiplied together. Thus, the probability of having a double-6 appear in a single throw of two dice is ⅙ times ⅙, or ¹⁄₃₆. The probability of getting three 6s in a single throw of three dice (or in three consecutive throws of a single die) is ⅙ times ⅙ times ⅙, or ¹⁄₂₁₆.

Rule 5

The probability that an event will happen, when the event's occurrence is dependent on another event taking place first, is equal to the probability of the first event times the probability of the second event's occurring after the first has already happened. This rule can be illustrated using a single die. If the object in two throws is to throw a 6, after first throwing either a 1 or a 2, the probability of doing so can be determined as follows: The probability of throwing either a 1 or 2 on the first throw is ⅙ + ⅙, or ⅓ (from Rule 3); and the probability of throwing a 6 on the second throw is ⅙. Therefore, the probability of first throwing a 1 or 2, then throwing a 6, is (⅙ + ⅙) × ⅙, or ¹⁄₁₈. Carrying this illustration a step further; if the object were to throw either a 4, 5, or 6 after first throwing either a 1, 2, or 3, the probability would be as follows:

$$(⅙ + ⅙ + ⅙) × (⅙ + ⅙ + ⅙) = ¼$$

Rule 6

The probability of having either one or the other of two events happening when the events are not mutually exclusive is equal to the sum of their separate probabilities, minus the probability of both happening at the same time. Assuming the objective were to throw either one or two sixes on one throw of two dice, the probability would be calculated as follows: the probability of throwing a 6 on one die is $\frac{1}{6}$, and the probability of throwing 6 on both dice at the same time is $\frac{1}{6} \times \frac{1}{6}$, or $\frac{1}{36}$ (from Rule 4). Thus, the probability of throwing either one or two sixes on a single roll is $(\frac{1}{6} + \frac{1}{6}) - (\frac{1}{6} \times \frac{1}{6})$, or $\frac{11}{36}$. This calculation can be verified by examining the previous table showing the 36 different combinations possible on two dice—exactly 11 of the combinations contain at least one six.

Probability, Poker, and Contracting

Many contractors refer to the contracting business as being much like a game of poker. And as pointed out earlier in the book, this comparison is valid in several respects:

1. There is an element of chance involved, but a much greater element of skill and judgment.
2. The skill in the game is based on judgments of the probabilities of different events occurring.
3. The competitive strength of a hand in poker, or of a bid in contracting, depends largely on the number of players (or bidders) in the game.

In both poker and contracting, even an elementary knowledge of the odds of the game will greatly improve the player's chances of winning.

There are 2,598,960 different five-card hands possible in poker. A hand consisting of two jacks, a 10, a 9, and an 8 has a .65 probability of winning in a three-handed game; in other words, there are approximately 1,690,000 possible hands that this one will beat, and about 900,000 hands that outrank it. In a 4-handed game there is a .52 chance of its winning; a 5-handed game reduces its chances to .42; in a 6-handed game, there is only a .34 probability of winning; a 7-handed game gives it a .28 chance; 8-handed, a .22 chance; 9-handed, a .18 chance; and 10-handed, a .15 probability.

To derive the maximum winnings from this hand, the strategy to be employed in the different situations may be considerably different—just as the best strategy to be followed by a contractor in bidding against six other competitors should be different from his strategy in competing against only two other competitors on a similar job.

The optimum or best strategy to be followed in both contracting and poker will be based to a large extent on the mathematical expectations associated with the particular competitive situation being encountered.

Mathematical Expectation

Mathematical expectation provides a means of striking a balance between the probability of an event's occurrence and the benefits to be gained by its occurrence.

Expectation is calculated by multiplying the probability by the amount to be gained. As such, mathematical expectation represents the average benefits that can be expected to result from a given situation over a long period of time, taking into account both the gains from successes and the losses from failures.

Referring again to the poker illustration from the preceding section—the hand holding a pair of jacks—we can visualize some of the peculiarities of mathematical expectation. Again, the probability of this pair of jacks being the winning hand depends on the number of players in the game, as follows:

Number of Hands	Probability of Holding Winning Hand
2	.81
3	.65
4	.52
5	.42
6	.34
7	.28
8	.22
9	.18
10	.15

A question, then, can be posed: Based on these probabilities of winning with the specified hand, which situation can be expected to yield the highest expected profit, assuming that each hand contributes $1.00 to the pot?

Calculating the mathematical expectation, or expected profit, for each case by multiplying the probability of winning by the amount of money in the pot that has been contributed by the other players gives the following results:

Number of Hands	Probability of Winning	Amount of Winnings	Profit Expectation
2	.81	$1.00	$0.81
3	.65	2.00	1.30
4	.52	3.00	1.56
5	.42	4.00	1.68
6	.34	5.00	1.70
7	.28	6.00	1.68
8	.22	7.00	1.54
9	.18	8.00	1.44
10	.15	9.00	1.35

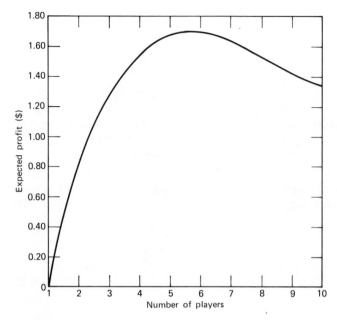

FIGURE 34.   Effect of number of players on expected profit.

This analysis of the expected profit associated with each situation leads to a somewhat surprising conclusion: The player holding the pair of jacks will win more money in the long run—his profit expectation will be highest, or optimum—in a six-handed game. The winnings will be greater than could be attained either by having a better chance of winning a smaller pot or by having less chance of winning a larger pot. Figure 34 shows the shape of the profit expectation curve for this example, beginning at 0 (representing a 100 percent chance of winning nothing when there is nothing in the pot except the player's own money), and increasing to its maximum value of $1.70 in a 6-handed game (a 34 percent chance of winning $5.00), then gradually decreasing to $1.35 in the 10-handed game ($9.00 in winnings, 15 percent of the time), as the decreased probability of winning more than offsets the larger amounts that might be won.

The general shape of the profit expectation curve leads directly to another interesting conclusion: The player will usually make more money by erring toward the high side of the "optimum" point on the expectation curve than by being the same distance toward the low side. In other words, profits are usually higher when the player takes a smaller chance of making a large gain than when he gambles on making many small gains.

The same line of reasoning carries over into many other fields, too. Often more money can be made by accepting a relatively low chance of winning a large amount than by having a high probability of gaining only a small amount.

By now the implications with respect to the contracting business should be clear: Higher overall profits may be possible by taking a few jobs with a high markup than by taking many jobs at a low markup.

Summary

Statistical techniques are valuable to the contractor in collecting, classifying, and analyzing numerical data and in using these data as a basis for predicting the outcome of future events. The first step in a statistical study is to accumulate and classify the desired information, usually in the form of a frequency distribution. From the frequency distribution a cumulative frequency distribution curve can be constructed, which can be used in assessing the likelihood of an event's future occurrence based on past performances. The theory of probability sets out specific principles by which the likelihood of an event's occurrence can be defined in numerical terms, thus minimizing the need for guesswork on the part of management. One of the main uses of probability theory is to develop the mathematical expectation associated with a variety of alternative actions or events by weighing both the probability of the event's occurrence and the potential benefits to be realized from its occurrence. Thus the types of situations most likely to yield maximum benefits, or profits, can be easily identified, and the effects of different actions can be evaluated.

14

Risk and Uncertainty

*Be not too presumptuously sure in any business; for
things of this world depend on such a train of unseen
chances that if it were in man's hands to set the tables,
still he would not be certain to win the game.*

GEORGE HERBERT

Contracting is an ideal business for those who enjoy taking chances. The
contractor is continually faced with a variety of situations involving many
unknown, unexpected, frequently undesirable, and often unpredictable factors.
These factors can be conveniently lumped together in the category of risk and
uncertainty.

An important requirement for successful management in the contracting field
is effective evaluation of the risks involved in the everyday business, followed by
sound decisions based on this evaluation, and appropriate action taken as a
result of these decisions.

While the terms risk and uncertainty perhaps imply different meanings, their
effects are similar. Risks generally refer to situations in which the distribution,
or probability of occurrence, of all possible outcomes is either known from past
experience or can be calculated with some degree of precision.

Uncertainty, on the other hand, covers situations which are of a relatively
unique nature and for which the probabilities cannot be calculated.

Risk, then, is simply a measurable uncertainty; uncertainty is a risk that
cannot be measured.

Taking risks is a necessary part of life. Business especially operates in an
environment of uncertainty, and an understanding of the "laws of chance" and
their effect on business is necessary to make intelligent management decisions
in such an environment. While chance and luck undoubtedly play some role in

business success, much of the skill in the business aspects of contracting is based on the ability of management to make realistic judgments of probabilities; the better prepared manager is better able to take advantage of "luck," and therefore appears to his competitors to be "lucky."

Risk and uncertainty cannot be avoided; management must decide in each situation whether there is more to be gained than lost in taking the necessary risks.

The Nature of Risk

Some economists believe that all true profit is related to risk, and if there were no risk involved in business, neither could business generate any profit. Compensation for risk and uncertainty was cited in Chapter 8 as one of the four basic elements making up profit.

Risks involve exposure to loss or injury. Some risks are well known enough that they can be insured against, with the contractor thereby assuming a direct cost and transferring the risk to an insurance company. Other risks, however, must be borne by the contractor.

The primary consideration in evaluating risks involves the potential returns or benefits that can be gained by taking the risks. In considering the risk of loss against the possibility of profit, the frequency of occurrence of the undesired events must first be estimated; next, the consequences of each risk on the business must be appraised; then, the cost of reducing risks by different degrees should be determined; and, finally, a judgment is necessary concerning whether the long-run profit expectation justifies the risks involved.

There are four different categories of risks involved in the contracting business; these risks may be of either a technical or economic nature.

The first category of risks are those inherent in the business. Simply to be engaged in the contracting business at all entails a great deal of risk, and these risks *must* be taken.

The second category of risks are those that *should* be taken—the risks assumed in pursuing opportunities for profits. These are generally economic risks, where the long-run effect will be beneficial, although numerous small losses may be incurred along the way.

A third category of risk is the risk too great to be taken. These risks are ususally of an economic nature, since technical risks can be allowed for and reflected in the bid price. But no bid, however large, can compensate for an owner who is unable to pay. And whenever the end result of taking such a risk could be complete financial disaster, the risk cannot be justified. Another example of a risk too great to be taken is that involving a contractor's taking a job that he is unable to perform.

The fourth risk category is the risk that is too good not to take; where the possibility of very high rewards outweighs the high risk involved.

The contractor's concern should not be with eliminating risks in his operations, but should be directed toward selecting the right risks to be taken.

The only way to eliminate risks is to do nothing—and in business this is actually the greatest, and most disastrous, risk of all.

Risks in Contracting

Scores of unexpected, or at least unpredictable, difficulties are apt to beset the contractor on any single job. And the longer the job, or the more separate activities involved in the job, the greater the likelihood of encountering problems. In minimizing the harmful effects of such difficulties, the contractor must first recognize the potential trouble spots; then, remedial action may be taken.

Risks as referred to in this chapter include all exposures to loss, whether brought about by nature, by shortcomings in the contractor's own organization, or by outside influences. No distinction is made between risk and uncertainty.

The following list includes a dozen of the major risks faced by the contractor:

1. Weather.
2. Unexpected job conditions.
3. Personnel problems.
4. Errors.
5. Delays.
6. Financial difficulties.
7. Strikes.
8. Faulty materials.
9. Faulty workmanship.
10. Operational problems.
11. Inadequate plans or specifications.
12. Disaster.

Weather

Problems caused by inclement weather are the least controllable of all the contractor's risks; they can be anticipated, but seldom controlled. Sufficient allowance, both in time and money, should be included in the price of the work. In determining the appropriate amount of such allowances, a study of historical data compiled by the U.S. Weather Bureau and other sources can be of considerable value.

Unexpected Job Conditions

Unexpected conditions refer to conditions encountered on the job that result in the actual amount of work required differing significantly from what was assumed when the job was estimated. Unexpected conditions are usually associated with underground work, particularly with the quantity and characteristics of rock and subsurface soil. Contracts should be studied closely by a competent lawyer to be sure that the contract documents clearly fix the responsibility for such unexpected conditions. If the contractor is legally responsible, time and money spent in additional site investigations may be necessary to reduce the possibility of future difficulties.

Personnel Problems

Personnel problems may arise in many different ways. The loss of key supervisory personnel can cripple a job, and this possibility must be at least considered. Difficulties may also be experienced in obtaining enough labor having the required skills. The local labor market should be investigated prior to submitting a bid, or the contractor may find himself unable to perform the work as scheduled, resulting in heavy penalties.

Errors

Whenever a dozen people are involved in as complex an operation as cost estimating or job scheduling, errors are likely to result. And the firm that makes the most, or the biggest, errors is the firm most likely to get the job. Major errors can be eliminated by careful checking; shortcut estimating methods are useful in detecting large errors or omissions in many types of jobs.

Delays

Delays in delivery of materials from suppliers' plants can tie up an entire job. The suppliers' records for promptness and dependability should therefore be a major consideration in the purchasing decision.

Financial Difficulties

Different types of financial problems may be encountered. Probably the most serious is the inability, unwillingness, or slowness of a client to pay; a review of

Dun & Bradstreet reports and credit ratings on a prospective client should disclose any major financial weaknesses, thereby minimizing such risks. Also, the retainage on some jobs may exceed the contractor's markup, thus requiring the contractor to finance part of the work out of pocket. Insufficient working capital in such cases could easily prove disastrous.

Strikes

Strikes and jurisdictional disputes have held up many jobs. Experience is the best guide to anticipating potential problem areas. The timing of labor negotiations and union contract expiration dates should always be considered when planning and scheduling a project. Also, labor tie-ups in suppliers' plants may cause excessive delays in deliveries.

Faulty Materials

The failure of materials to meet specifications is always a possibility to be considered. Experience with the particular materials to be used offers the best guide to their acceptability. The "or equal" clauses found in many contracts should be cautiously interpreted to avoid future inconveniences and possible increased costs.

Faulty Work

Poor workmanship is doubly expensive, resulting both in increased job costs and in a damaged reputation for quality work. Problems of this type can be largely avoided through careful and capable supervision, by maintaining a few key craftsmen on the payroll, and by being as selective as possible in hiring practices.

Operating Problems

Labor seldom performs exactly as expected. Should performance lag too far behind what was originally estimated, additional expenditures must sometimes be made in increased supervision. Damage to equipment through overloading or poor maintenance and loss of tools and materials through theft or vandalism are common occurrences that must be anticipated.

Inadequate Plans and Specifications

Contract documents are sometimes vague, and much money has been lost because of the contractor's failure to clarify the exact responsibility of each party. Responsibility for correcting errors or defects in the original plans and specifications should always be spelled out in advance.

Disaster

Disaster refers to a sudden and extraordinary misfortune, causing great destruction of life or property, such as fires, floods, and tornadoes. Insurance can cover many disasters; if the effect of a major disaster would be to completely ruin the company, a lack of insurance is a poor way to economize.

Reducing the Effect of Risks

Risks can never be avoided or eliminated, and even minimizing risks is not always desirable. But several methods can be employed in meeting risks and uncertainty to reduce their possible harmful effects. The methods fall generally into six broad but overlapping categories:

1. Consolidation.
2. Specialization.
3. Control.
4. Prediction.
5. Diffusion.
6. Selection.

Consolidation

The degree of uncertainty is much less in groups of cases than in single instances. Consolidation refers to the grouping of risks to take advantage of the "law of large numbers." In effect, by grouping a large number of risks the contractor is performing the function of an insurance company. If he calculates his risks carefully enough, and encounters the same types of risks frequently, he will be able to cover himself adequately.

Specialization

The specialization principle is similar to the consolidation principle, except that the responsibility for assuming risks is delegated to an individual or a firm that specializes in handling risks, such as an insurance company. Thus the insurance company actually consolidates risks by assuming risks for a large number of different concerns and, by doing so, minimizes its own risks. The contractor, instead of assuming a risk, substitutes a direct cost—the insurance premium—for his risk.

Control

Control over an uncertain future would be an effective way to reduce the harmful effect of risks and a greater ability to control the course of future events can be achieved. Additional safety measures, for example, can provide effective control over certain types of accident risks commonly encountered on most jobs.

Prediction

Increased powers of prediction, thereby enabling the contractor to foresee possible future difficulties, will allow him to compensate more precisely in his pricing structure for the risks to be encountered. Continual surveillance of pertinent business information, including government and trade publications and private services such as Dodge Reports, can keep the contractor well informed on present and probable future developments in his field.

Diffusion

The diffusion principle refers to the spreading out of the harmful effects of risks, should the unfortunate event actually happen. The idea behind the diffusion principle is that the effect of a $1000 loss on each of a hundred individuals is less harmful than a $100,000 loss suffered by a single individual. This principle is closely related to the consolidation and specialization principles. A practical application of the diffusion principle would be in avoiding excessively large investments in any one project; if one job should prove disastrous, then the losses could be absorbed by income from other jobs.

Selection

Operations can frequently be directed along lines involving a minimum of risk and uncertainty, thus avoiding the less desirable areas. Business risks can be avoided only by avoiding business, but equally profitable opportunities can sometimes be found with substantially different risk levels involved.

Risk Evaluation Through Simulation

Simulation involves the use of a mathematical or analytical model to study objectively the effects of risk factors on the proposed project. Seven steps are included in the simulation process:

1. Identify the risks.
2. Estimate the probability of their occurrence.
3. Assess their consequences.
4. Prepare a project schedule.
5. Formulate a model.
6. Feed in the variable factors.
7. Interpret the results.

Identifying the Risks

A dozen of the most common types of risks facing the contractor were discussed in an earlier section. First step in the simulation process is to determine exactly which risks are apt to apply to the project at hand.

Estimating the Probability of Occurrence

Estimating the probability of occurrence of most risks requires a combination of factual historic data and informed intuitive judgment. Long-range weather forecasts are usually available and can be used as a basis for predicting the chances of having delays caused by inclement weather. Strikes can often be anticipated from past experience, and a study of past records can provide insights into the likelihood of encountering many other types of difficulties. Additional preliminary work in this stage can often pay off handsomely in the long run.

Assessing the Consequences of Risks

The harmful effect of each risk—should it come to pass—on the project should next be examined. A certain level of rainfall, for example, might halt work for one day during the early stages of a project but have less effect later in the project. A greater amount of rainfall during a short period of time might result in several days lost on certain parts of the job, thereby requiring schedule revisions throughout. Again the contractor's experience will be the best guide to the effect of risks on his own operations.

The Project Schedule

Preparation of the complete project schedule is the next step. This schedule should clearly define the manpower and equipment requirements for each phase of the work, specify the necessary delivery dates for materials and supplies, and delineate the role of subcontractors in the project.

Formulating an Analytical Model

The manpower and equipment requirements defined in the preceding step can be used in preparing a model of the system. A critical path network diagram showing the interrelationships among all the different tasks involved in the work makes an ideal model for determining the effect of risks on the overall project.

Applying the Risk Factors

Once the model has been established, the effect of different risk factors on the project can be calculated. While many simulation problems are best solved by means of an electronic computer, useful results can generally be obtained by clerks and desk calculators on all but extremely large and complex projects. The effect of weather, for example, can be calculated by assuming 1 in. of rainfall during a specified time period, identifying its effect on different tasks, making the appropriate schedule changes throughout, and calculating the effect of these changes on the total project, both in terms of time and cost.

Interpreting the Results

By repeating the preceding calculations under different assumptions, a relatively clear picture of what might be expected can be obtained. For example, if the effect of no rain, 1 in. of rain, and 2 in. of rain were evaluated, the resultant figures considered in view of their relative frequency of occurrence could provide a realistic basis for estimating the project completion time. Similarly, the effect of other variable factors, along with the probability of their happening, should lead to greatly improved project planning and control.

The Monte Carlo Technique

Many important business decisions are decided in the face of uncertainty on the toss of a coin—"heads we will, tails we won't." This method, lacking the degree of sophistication required by professional statisticians, has been refined somewhat and renamed the Monte Carlo Technique.

The Monte Carlo technique involves essentially using a 10-faced coin or a 10-sided die, or a table of random numbers. Random numbers are just what their name implies—numbers chosen completely at random. Table 22 contains

TABLE 22. Random Numbers

69709	31742	48961	74344	07439
31354	16857	32615	87525	21156
64920	46823	87511	69439	80535
49861	75138	72533	27249	70024
62179	18960	89700	25890	86207
56631	90374	22806	08413	73532
20763	80505	08562	43957	18069
57186	90911	78493	42854	26194
61969	34871	69133	42605	67681
65627	97054	93012	31471	98916
65800	78081	93594	52367	01273
80451	32603	15196	00311	64587
55698	82927	28702	87028	78345
45895	73148	90942	43952	54437
48769	27250	84342	39563	52407
32699	44701	12508	00726	85370
81162	39480	83157	53508	46157
49827	01527	32661	44259	18560
02353	70569	82385	26948	53534
17907	70294	10976	39209	47846

500 random numbers, compiled by taking the last three digits of individual state populations from 1880 to 1950. In this table there is approximately an equal number of each digit, 0 through 9. And since the numbers were chosen completely at random, a random selection can be obtained by starting at any point in the table and moving in any direction.

Entire books containing nothing but random numbers generated by electronic computers have been compiled for statisticians' use. Table 22 should be adequate, however, for finding experimental solutions to contractors' problems involving uncertainty.

The Monte Carlo technique can be used in a number of different applications to study the effect of random factors on a project schedule.

For example, consider a problem involving queueing theory where trucks are expected to arrive during a certain time interval, but where their exact arrival times are unknown. Using the Monte Carlo technique, an adequate solution to the problem can be found by assuming the truck arrival times to be distributed completely at random, estimating their times from the random number table. By working the problem out with several different sets of random numbers, a solution can be obtained that will reflect actual conditions as closely as possible in the absence of more specific information.

The following table shows how the truck arrival times could be estimated, using random numbers from Table 22.

10-Minute Time Interval	Random Numbers			Estimated Arrival Times		
	Trial 1	Trial 2	Trial 3	Trial 1	Trial 2	Trial 3
9:00–9:10	6	1	9	9:06	9:01	9:09
9:11–9:20	9	7	6	9:19	9:17	9:16
9:21–9:30	7	4	1	9:27	9:24	9:21
9:31–9:40	0	2	7	9:40	9:32	9:37
9:41–9:50	9	4	4	9:49	9:44	9:44
9:51–10:00	3	8	3	9:53	9:58	9:53

The assumption has been made that one truck will arrive sometime during each 10-minute interval. Therefore, the 10-minute intervals were set up first. Then an exact arrival time during each interval was estimated by taking random numbers from Table 22. Calculations can then be made regarding the waiting time to be encountered in each case, and an average of the results of each trial calculation should yield an answer precise enough for decision-making purposes.

Statistical Measurements

Very few exact measurements can be made regarding either time or money. At best, an average value is possible; and usually some indication of minimum and maximum values can be established.

For example, the actual time required to complete a specific task may vary anywhere between 90 and 110 percent of its usual completion time, while another task may vary between 80 and 120 percent of the most commonly encountered time. Some idea of the frequency with which each outcome occurs is therefore useful in developing accurate estimates of project completion time.

Expected Time

By making three time estimates for a given task—representing the most optimistic, most pessimistic, and most likely times—the "expected" or average time can be approximated from the following formula:

$$t_e = \frac{a + 4m + b}{6}$$

where t_e is the expected, or average time; a is the shortest possible time; b the longest conceivable time under normal conditions (but disregarding the possibility of a major disaster); and m, the most likely, or most often occurring time.

Figure 35 shows some typical frequency distribution curves which could apply equally well to either time or cost measurements. Parts *a, c,* and *e*

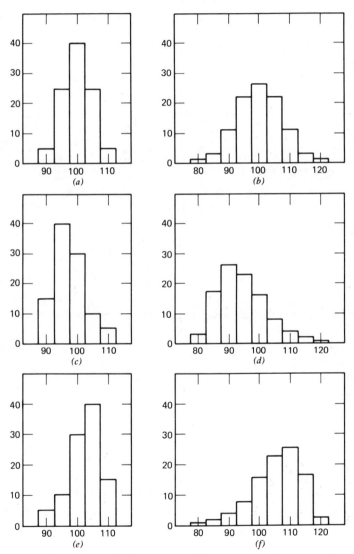

FIGURE 35. Different types of frequency distributions.

represent frequency distributions ranging between values of 90 and 110; the distribution in *b, d,* and *f* vary from 80 to 120 units. The units could be in terms of either time or money.

In *a* and *b,* the most likely time is 100; in *c,* the most likely time is 95; in *d,* 90; in *e,* 105; and *f,* 110. Obviously, to estimate these jobs at their most likely time could lead to serious problems, for the most likely times correspond to the average times only in *a* and *b.* Table 23 summarizes the data for each of these frequency distribution curves and shows how the actual average times compare with the approximation calculated from the preceding formula.

In Figure 35 *c,* for example, the actual average is 97.5, and the approximate expected value is 96.7. In Figure 35 *d,* the average is 94.4 against the calculated expected value of 93.3. The approximation of the expected value, as calculated from the formula, gives a result within about 2 percent of the actual average—close enough for all practical purposes—without requiring the construction of a frequency distribution curve.

The formula can be used safely with most time or cost data. For example, if the most likely total cost of a certain type of brickwork were $150 per thousand bricks, but might vary between $125 and $200 under commonly encountered

TABLE 23.   Statistical Measurement of Different Frequency Distributions

Value	A	B	C	D	E	F
			Distributions			
75	0	0	0	0	0	0
80	0	1	0	3	0	1
85	0	3	0	17	0	2
90	5	11	15	26	5	4
95	25	22	40	23	10	8
100	40	26	30	16	30	16
105	25	22	10	8	40	23
110	5	11	5	4	15	26
115	0	3	0	2	0	17
120	0	1	0	1	0	3
125	0	0	0	0	0	0
Minimum	90	80	90	80	90	80
Maximum	110	120	110	120	110	120
Most likely	100	100	95	90	105	110
Average	100	100	97.5	94.4	102.5	105.6
Expected	100	100	96.7	93.3	103.3	106.7
Variance	11.1	44.5	11.1	44.5	11.1	44.5

conditions, the correct cost for estimating purposes would be as follows:

$$\frac{125 + 4 \times 150 + 200}{6} = \$154$$

Such an estimate considers the possibility of both better and worse performance than is normally expected.

Variance

The statisitcal term *variance* refers to the reliability of estimates. If the variance is high, the degree of uncertainty associated with the work is high; a low variance reflects good accuracy. Variance is a function of the extreme range of values—the minimum and the maximum—and can be calculated from the following formula:

$$V = \left(\frac{b - a}{6}\right)^2$$

The variances for each of the frequency distributions of Figure 35 are also summarized in Table 23 and indicate simply that estimates based on distributions B, D, and F are less reliable than estimates based on distributions A, C, and E, which could be observed without any calculation at all. These measures of variances are sometimes used in PERT calculations; however, such statistical hanky-panky has little significance for most contractors in the construction field.

Responsibility for Unexpected Job Conditions

The question of who pays for the unexpected in construction has plagued contractors for many years and "changed conditions" have long been a cause of disputes among contractors, clients, and consulting engineers and architects. In today's construction market, unexpected job conditions are generally considered to be the owner's responsibility, and the contractor is very rarely held responsible.

Changed conditions usually are encountered in excavation or foundation work when the quantities, dimensions, or classes of work actually encountered differ substantially from the work scope shown in the contract documents. The main argument, then, centers around whether the conditions encountered on the job should actually have been unexpected. Frequently the owner reasons that the costs should have been anticipated by the contractor, while the contractor maintains that there was no reason to expect such conditions at the time the job was let. Assuming that a mutually satisfactory agreement is reached regarding responsibility for the changed conditions, the only major question to be resolved concerns the price of the additional work.

A thorough investigation of the proposed construction site is the surest way to minimize the possibility of problems arising from unexpected conditions. A second method, commonly used but less satisfactory, is to include unit prices in the contract.

Site Investigation

A site investigation, including test borings and soils analysis, could undoubtedly reveal many conditions that would later qualify as "unexpected." The contractor should, at least, visit the site to inform himself of the surface physical conditions and gather any available information from neighboring locations and from municipal or state records. Nearby outcrops, borrow pits, and other visual factors, as well as interviews with local residents and contractors, will sometimes uncover many factors of considerable significance. Interpreting such information is the contractor's rsponsibility; but if truly unexpected conditions should arise later, the owner should pay. Again the question revolves around the definition of "unexpected"; the contract documents must make the definition clear. To search for legal loopholes in the contract after the conditions have already been encountered is unwise.

Unit Price Contracts

Some owners feel that unit price contracts will eliminate any problems concerning unexpected job conditions. Serious difficulties can still arise, however, when the quantities encountered differ substantially from the quantities anticipated. For example, where a $50 per cu yd unit price estimate for rock excavation might be tolerable if only 100 cu yd of rock was expected, should several thousand cubic yards actually be encountered in the job, the owner would be scrambling for the legal loopholes. Careful wording and concise definitions in the contract are necessary to avoid major conflicts.

Responsibility for Other Unexpected Conditions

In general, delays beyond the contractor's control are compensated for by allowing him extra time to complete the project—a poor substitute for money. Some provision should be made in the contract that would reimburse the contractor for actual damages caused by the owner or his engineers or architects, such as by their failure to secure the necessary rights-of-way, by their delaying the work pending changes or revisions, or through faulty plans or specifications.

Summary

Risk and uncertainty are an integral part of business and are especially important in the contracting field. The contractor's objective should be to choose the right risks—the ones that are most strongly outweighed by the potential profits. While the risks themselves cannot be eliminated, their harmful effects can be minimized through consolidation, specialization, control, prediction, diffusion, or selection. Simulation, by means of a mathematical or analytical model, is a useful technique for examining the effect of risks on a project schedule; the Monte Carlo technique also has some useful applications in combating uncertainty; and the degree of uncertainty in many situations involving time or money can often be reduced, or at least compensated for, through the use of simple statistical methods. Unexpected job conditions, calling for changes and revisions in construction schedules, frequently result in disputes between contractors and owners; consequently, possible changes should be anticipated as much as possible prior to submitting bids, and definitions of and responsibility for the unexpected in construction should be clearly spelled out in the contract documents.

15

Information Management

A business man's judgment is
no better than his information.

R. P. LAMONT

Information—like garbage—should not be collected until arrangements have been made for its disposition. Unless there is some place to put it after it has been collected, some way to find it when it is needed, and some useful purpose for it after it has been found, information has little value to anyone.

One of the basic principles of physics states that the creation of order will not occur by itself; energy must be expended to form an ordered system. This principle most certainly applies to the contractor's information system. Creating order out of the masses of available data requires considerable effort and expenditure of energy. Once an ordered system has been created, though, maintaining it is an easy routine task readily handled by clerical personnel.

Why an Information System?

The purpose of the information system described in this chapter is to provide the contractor with the types of competitive information that will enable him to achieve his strategic goals. Having the right information available to the right person at the right time will provide the objective basis needed for informed decisions.

A word of caution is in order here regarding the collection of information. There are only two good reasons for collecting and/or saving information:

1. The information is required by law. This type of information includes payroll and accounting records required by federal, state, and local agencies.

184

2. The information is useful to management in its business planning and operations. This type of data—specifically, that which deals with the competitive aspects of the business—is the main concern of this chapter.

Accounting records must be kept in a certain way and are the domain of the firm's accountant or bookkeeper. Competitive information for use by top management should be kept in an easily maintainable and readily accessible manner for management's immediate reference and use.

Requirements of an Information System

Any information system, to be effective, must be capable of handling whatever kinds of data will be useful in the management decision-making process. The data contained in the information system must be readily available to the decision maker whenever it is needed and in a form that can be easily interpreted and used. The ideal information system consists of human judgment and imagination, supported by relevant data; management supplies the inspiration, while the system "hardware" provides access to the relevant facts.

The information system should be flexible enough to allow the manager to introduce as many facts as will be useful to him and to retrieve these facts quickly and conveniently. The information system, therefore, is simply a fact retriever; it does not supply any judgments. Management gets paid for its judgment skills.

The information system only makes available the facts that form the basis for management decisions, with the manager himself adding values to the facts at his own discretion and in terms of his own judgment, experience, and intuition.

By having large quantities of organized data easily accessible, the manager will be able to spend more productively his time in applying his judgment to the data rather than searching for the data. He will also have a factual and objective means of justifying his decisions once they are made.

Computerized Versus Manual Information Retrieval

The information system described here will undoubtedly be dismissed as old-fashioned or outmoded by many contractors who are currently employing computers in their operations. For these contractors, this information system may be antiquated, and the feasibility of incorporating the information and analysis procedures into their computer system should certainly be investigated, both in terms of cost and convenience.

For contractors who do not have regular and easy access to a computer, though, the information system covered in this chapter should provide a practical and convenient substitute. In fact, according to several knowledgeable computer systems analysts, the edge-notched data cards may, for the applications described here, be superior to a computer.

The manager can retrieve and analyze the data from the data cards far quicker than he can walk down the hall, explain to his computer expert what he needs, submit the appropriate requests to the computer, and obtain the desired results. The only real time saving offered by a computer is in the actual arithmetic computations. However, these computations are simple, and, when accomplished manually, may give the manager special insights and a better personal feel for his market and competition.

The Contractor's Information System

Very large contractors, bidding hundreds of jobs each year throughout the country, may be able to justify the use of a computer-based data system for storing and retrieving the competitive information to be employed in their bidding strategy.

At least 95 percent of all contractors, though, can get by with the system described in this chapter. It employs a minimum of "hardware" and a maximum of "brainware" and can compete effectively with any computer in the world.

The information system hardware consists of just three elements, pictured in Figure 36:

1. A deck of *cards* with holes drilled around the edges, for recording the data. One card will be used for each job to be recorded, so a package of several hundred cards will last the typical contractor a long time. The cards range in size from about 3 × 5-in. up to 8 × 10-in., and are locally available from office supply stores, college book stores, and office machine and equipment dealers. The card size and number of holes determine the amount of data that can be stored; either 3 × 5-in or 5 × 7-in. cards are usually adequate.

2. A *hand punch* for notching the cards to indicate the data recorded on each. These punches come in several styles, from the same sources that supply the cards. In the absence of an "official" edge-notcher, an ordinary pair of scissors can be used.

3. A *sorting rod* for retrieving the cards containing the desired information. An icepick, awl, knitting needle, or any other rod long and thin and rigid enough to slip through the holes on the cards and support 100 or so of them at a time will suffice.

The whole system will cost less than $50. The cards can be kept in a regular desk-top file box, a cigar box, or any other handy container.

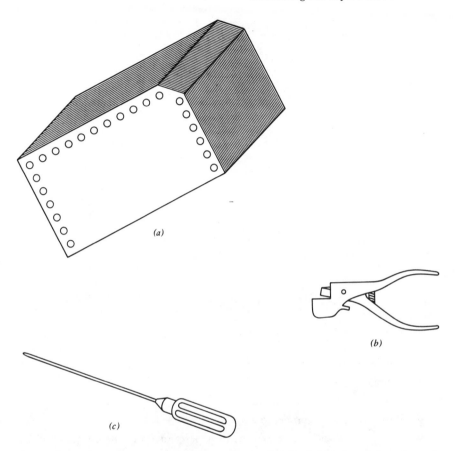

(a)

(b)

(c)

FIGURE 36. Information system hardware. (a) A deck of drilled cards (one for each job to be recorded) on which to record the relevant data; (b) a hand punch for notching the cards to indicate the data recorded on each; (c) a sorting rod for retrieving the cards containing the desired data.

With the information system hardware in hand, the next step is to record the information that will be used in analyzing the competitive information and developing the bidding strategy.

Recording the Information

The information to be stored must first be recorded on the data cards. Since the source of information is generally the bid tabulations on past jobs, the usual way to record the information is simply to type the relevant information from a

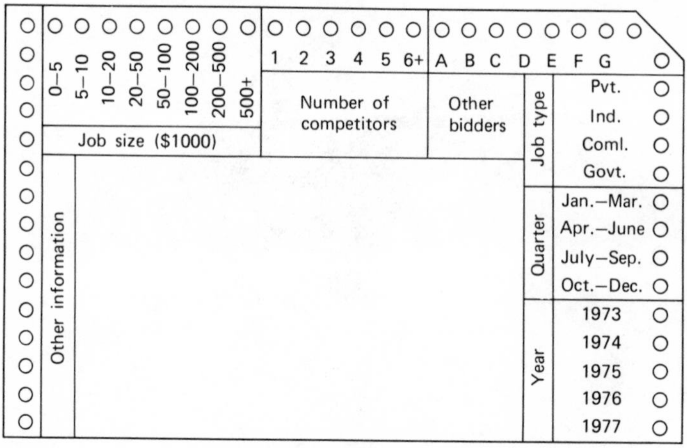

	0–5	5–10	10–20	20–50	50–100	100–200	200–500	500+	1 2 3 4 5 6+	A B C D E F G	Job type		Year/Quarter	
									Number of competitors	Other bidders		Pvt.	
												Ind.	
	Job size ($1000)											Coml.	
												Govt.	
												Jan.–Mar.	
	Other information										Quarter	Apr.–June	
												July–Sep.	
												Oct.–Dec.	
												1973	
												1974	
											Year	1975	
												1976	
												1977	

FIGURE 37.   Data card.

single job onto a card. Each card, then, will contain the bid tabulation for a specific job.

Typically, the information to be recorded for future use includes these items:

- Job description—type of work, location, date, owner, and so on.
- Estimated direct job cost.
- Names of competitors bidding the job.
- Bid prices for all competitors.
- Competitors' bid prices expressed as percentages of the estimated direct job cost.
- Any other information that might be useful later.

The information recorded on the data card is nothing more than a complete bid tabulation for a specific job. Figure 37 shows a convenient 3 × 5-in. data card that will be adequate for most contractors. Hundred of different cards are commercially available in different sizes and configurations. Most data cards will have the top right corner cut off so that the cards can be lined up properly.

Coding the Information

Coding the information recorded on the data cards is accomplished by assigning appropriate significance to the holes around the card edges. Each hole used represents a certain specific piece of information.

There are many ways that the information on a data card can be coded, and enormous amounts of information can be accomodated by employing different coding techniques.

The contractor, though, is not faced with enormous quantities of data, so his coding system can be quite simple.

The data card shown in Figure 37 has 51 holes around its edges: 23 across the top and 14 along each side. Should 51 bits of information not be enough, another 22 holes could be drilled along the bottom or a larger card could be used.

Starting with a look at the information to be recorded, it is not too difficult to figure how many holes will be needed. For example, the contractor may, on examination of his past jobs, decide that the following classifications are appropriate for his operation:

- Job size—big or little.
- Number of competitors—few or many.

In this oversimplified case, six jobs might have the following characteristics:

Job Number	Job Size	Number of Competitors
1	small	few
2	big	many
3	big	few
4	small	many
5	small	few
6	big	many

These jobs characteristics would be coded on the data cards as shown in Figure 38. The job characteristics to be recorded are recorded on the cards by cutting a slot from the card edge to the hole as shown in the illustration.

Job 1, the first card in Figure 38, has two slots cut in it: the first indicates a small job and the second shows that there were few bidders on the job. Jobs 2 to 6 are coded in the same way on the remaining five cards.

Retrieving the Information

The final step in any information handling system involves getting the information that was put into the system back out of the system.

In sorting the cards, a sorting rod is inserted through the holes representing the data to be retrieved. Then, when the rod is lifted and shaken, only the cards

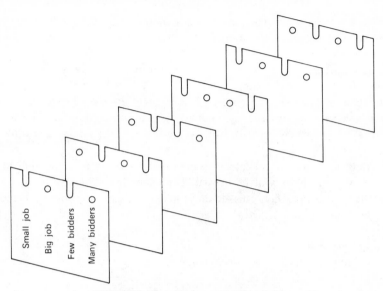

FIGURE 38. Coding the information.

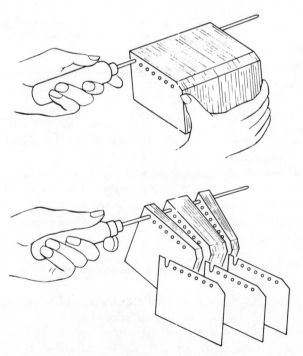

FIGURE 39. Sorting the data cards.

that have been notched will drop out. The rest will be retained on the rod. These notched cards that dropped out contain the desired data. Figure 39 shows how this works.

In this case, if the sorting rod is inserted through the first hole ("small job") (refer back to Figure 38), cards 1, 4, and 5 will drop out, with cards 2, 3, and 6 still remaining on the rod. All of the cards that drop out have the desired job characteristic that was being sought—they are all small jobs.

If the information wanted were all small jobs bid against few competitors, a second sort would be made of just the cards that dropped out on the first sort.

The first sorting retrieved all of the small jobs; the second sorting then found, from among the small jobs only, those bid against few competitors.

Example of the Information System

The following three tables (Tables 24, 25, and 26) and six figures (Figures 40 through 45) show how the information system might be applied.

TABLE 24.   Bid Tabulations on Jobs 1 and 2

Bid Tabulation

Job Number: 1   Date: April 26, 1972
Estimated Direct Cost: $19,000   Job Type: Residential

Rank	Contractors Bidding	Amount Bid	Percentage Markup
1	Lester P. Whipsnade Corp.	$22,800	20.0%
2	Underbid Building Co.	23,100	21.6
3	Aardvark Bidding Co.	23,700	24.8
4	Gandy Brothers	26,400	39.0

Bid Tabulation

Job Number:2   Date: May 3, 1972
Estimated Direct Cost: $43,000   Job Type: Commercial

Rank	Contractors Bidding	Amount Bid	Percentage Markup
1	Underbid Building Co.	$46,400	8.0%
2	Cheapjack Construction Co.	49,500	15.2
3	Aardvark Bidding Co.	53,700	25.0
4	F. B. Knight Contractors	55,800	29.9
5	Lester P. Whipsnade Corp.	61,700	43.5

TABLE 25. Bid Tabulations on Jobs 3 and 4

Bid Tabulation

Job Number: 3   Date: May 7, 1972
Estimated Direct Cost: $76,000   Job Type: Public

Rank	Contractors Bidding	Amount Bid	Percentage Markup
1	Emmett Swink & Sons	$87,400	15.0%
2	Aardvark Bidding Co.	95,000	25.0

Bid Tabulation

Job Number: 4   Date: May 10, 1972
Estimated Direct Cost: $15,000   Job Type: Residential

Rank	Contractors Bidding	Amount Bid	Percentage Markup
1	Cheapjack Construction Co.	$16,500	10.0%
2	Gandy Brothers	17,400	16.0
3	Aardvark Bidding Co.	18,800	25.4
4	Emmett Swink & Sons	19,600	30.7
5	R. Dink Contracting Co.	23,500	56.7
6	Lester P. Whipsnade Corp.	26,700	78.1
7	F. B. Knight Contractors	29,200	94.7

The tables summarize the bid tabulations on six different jobs. The figures show how the data would be recorded for each job, on a 3 × 5-in. data card containing 51 holes. Here, only 34 of the holes are used which represent the following:

- Job size (8 categories).
- Number of competitors (6 categories).
- Other bidders (competitors) (7 categories)
- Job type (4 categories)
- Quarter when job was bid (4 categories).
- Year (5 categories).

The data cards representing all six jobs are shown together in Figure 46. Here it can be seen what can be accomplished by sorting the cards. While the six

TABLE 26.   Bid Tabulations on Jobs 5 and 6

Bid Tabulation

Job Number: 5   Date: May 14, 1972
Estimated Direct Cost: $22,000   Job Type: Commercial

Rank	Contractors Bidding	Amount Bid	Percentage Markup
1	Aardvark Bidding Co.	$27,500	25.0%
2	Gandy Brothers	30,800	40.0
3	F. T. Coop Builders	32,700	48.6

Bid Tabulation

Job Number: 6   Date: May 21, 1972
Estimated Direct Cost: $62,000   Job Type: Industrial

Rank	Contractors Bidding	Amount Bid	Percentage Markup
1	F. B. Knight Contractors	$63,200	2.0%
2	Cheapjack Construction Co.	66,000	7.6
3	F. T. Coop Builders	71,400	15.2
4	Aardvark Bidding Co.	77,500	25.0
5	Emmett Swink & Sons	83,300	34.4
6	R. Dink Contracting Co.	94,800	53.0

cards in the illustration pose no problem for visual sorting, this system can just as quickly pull out all jobs having any one desired characteristic from among 100 or more cards.

By inserting the sorting rod in the hole representing the third category of job sizes, for example, all jobs falling in the $10,000 to $20,000 range will drop out. Similarly, all jobs in that size range bid against three competitors can be pulled out on a second sort. If every job bid against the Cheapjack Construction Company were wanted for examination, that too could be easily accomplished. Sorting could be made to determine seasonal variations, to identify commerical clients, or to find anything else recorded on the cards. When new data become available, a whole year's accumulation of data cards could be pulled out, and whatever analyses are conducted could be revised or updated. There is virtually no limit to the combinations of data that can easily be sorted out, and the cards need not be kept in any particular sequence.

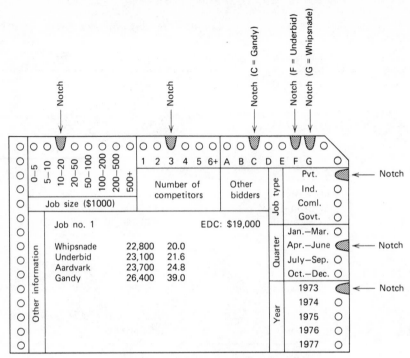

FIGURE 40. Data card coded for job 1.

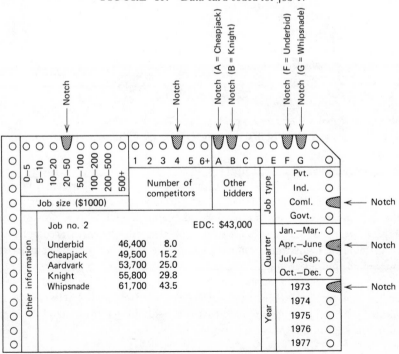

FIGURE 41. Data card coded for job 2.

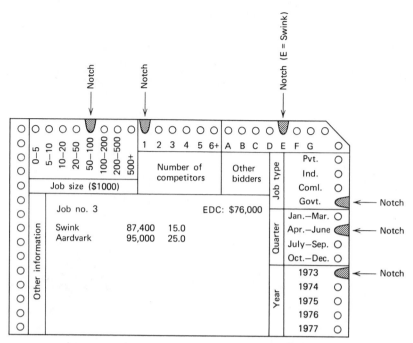

FIGURE 42. Data card coded for job 3.

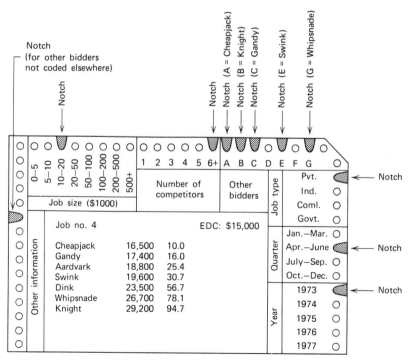

FIGURE 43. Data card coded for job 4.

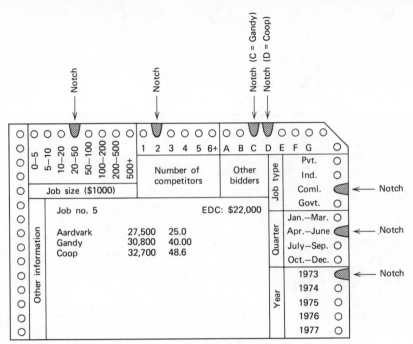

FIGURE 44. Data card coded for job 5.

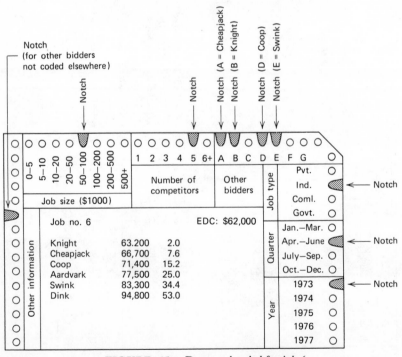

FIGURE 45. Data card coded for job 6.

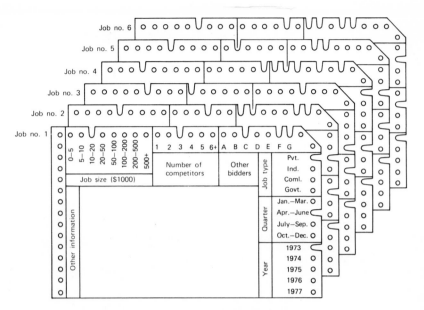

FIGURE 46. Data cards coded for jobs 1 to 6.

Summary

A good information system will provide the contractor with whatever competitive information he needs, in whatever combinations he needs it, whenever he needs it. The information system described in this chapter can be put together for less than $50 and be maintained and updated by clerical personnel. Even so, it will quickly and conveniently provide the specific information needed for analyzing jobs and competitors in developing an effective bidding strategy.

16

Analyzing the Competition

*Before you organize you ought to analyze and see
what the elements of business are.*

GERARD SWOPE

The analysis of past bids can provide much useful information for the contractor in making the best possible decisions on present matters in the face of uncertainty; uncertainty can almost always be lessened, and often eliminated, by careful analysis.

The purpose of bid analysis is to identify significant relationships among the many variable factors involved in competitive bidding, so that these relationships can be used to predict the course of future events. Many of these relationships are recognized intuitively by experienced contractors, but other relationships that are generally believed or assumed to be correct may actually have no basis in fact. Bid analysis can quickly separate truth from fiction.

Many significant relationships can be determined through analysis. The following are particularly important and are covered in detail in the subsequent sections:

1. Number of bidders per job.
2. Distribution of competitors' bids.
3. Distribution of low bids.
4. Effect of the number of bidders on the low bid.
5. The spread.

Most of the information presented in this chapter is taken from actual records of contracting firms, primarily in the building, heavy, and mechanical fields. As such, the data are factually accurate; they are not, however, necessarily representative of any firm other than those whose records were

198

analyzed. The relationships presented in the following sections, therefore, are intended only as examples and must be interpreted with caution.

Such information is easy to develop for any contractor who maintains adequat records and who is willing to spend the few hours required to study and analyze his records and to establish the meaningful relationships.

Information Requirements and Sources

The information requirements for bid analysis are simple and straightforward, although obtaining some of the information is sometimes difficult.

Essentially, only three general categories of information are required:

1. General information concerning specific jobs, including the type of job, location, date, unusual conditions, and so on.
2. The company's estimate of the direct costs associated with each job; this figure should include job overheads, but not general overheads.
3. Tabulations of competitors' bids on all jobs, especially those for which estimates were prepared.

The first two items—job characteristics and estimated job costs—can be obtained from the company's own records. Only the third item—competitors' names and bids—is likely to cause trouble.

On public works projects, and on many jobs carried out by private owners, public bid openings are common. In these cases the names and bids of all competitors are made known, and no information problems arise. Bid tabulations are also published in construction trade journals and by organizations such as F. W. Dodge.

Some private owners, however, are reluctant to reveal the bids submitted on their projects, perhaps feeling that such disclosure would decrease their subsequent bargaining power. Even so, most contractors are able to compare notes with those of their associates, and "secret" bids do not generally remain secret for long.

The contractor should carefully compile and retain all such information. Every competitor's bid on every job for which a detailed cost estimate is available will be extremely valuable. Table 27 shows examples of two actual bid tabulations reported in the construction press.

Effect of Job Size on Number of Bidders

The average number of bidders on a job will depend both on the job characteristics and the general competitive situation within the industry.

TABLE 27.   Examples of Published Bid Tabulations[a]

Church-School Addition–Omaha, Nebraska. St. William's Church, 1544 North 18th Street; Owner, c/o Architect: Schluck & Schluck, 626 South 12th Street, Omaha. Billy Club Construction Co., Winslow, Nebr. apparent low bidder at $138,844. Mechanical and electrical work in general contract; kitchen equipment bids not included. One-story semibasement type structure; concrete and steel frame; concrete exterior walls; parking deck; 64 ft. × 84 ft. building contains gym, kitchen, toilet, and storage rooms. Hot water heating system and air conditioning next low bidders:

Otto Park & Co.	$139,000
Jack Handle Construction Co.	141,000
Ray Gunn & Son	142,500
Harold Square Construction, Inc.	142,650
Charlie Horse Constructors	144,950
Rip Chord & Associates	148,650

School Addition–North Fork, Nebraska. P.S. No. 110, c/o Architect: Tom, Tom & Associates, P. O. Box 9232, North Fork, Nebr. Harry Legs & Son, North Fork, apparent low bidder on general contract at $104,449; Aardvark Engineering, Inc., North Fork, apparent low bidder on mechanical work at $17,397; Zap Electric Co., North Fork, apparent low bidder on electrical work at $5,832. Next low bidders:

General
Dewey, Cheatum, & Howe	$110,470
Larcen & Steele	113,689

Mechanical:
Boiler Room Engineering Co.	$17,433
Attic Engineers & Contractors	19,410

Electrical:
Hi Voltage Electric Co.	$5,900
Watt Electrical Contractors	7,325
Short Circuit Company	9,246

[a] All names are fictitious.

Logically, large jobs, offering a potentially greater profit opportunity, should attract more bidders than small jobs. However, as the size of job increases the number of contractors qualified for the work is likely to decrease, as bonding or credit requirements eliminate some of the marginal operators.

A sample of 100 jobs ranging in size from about $10,000 up to $70,000,000 was drawn from published sources over the past few years. Over all, an average of seven bids was submitted on each job. The average number of bidders varied with the job size as follows:

Size of Job	Average Number of Bidders
Under $50,000	4.8
$50,000 to $100,000	7.1
$100,000 to $500,000	5.5
$500,000 to $1,000,000	7.3
$1,000,000 to $5,000,000	8.3
$5,000,000 to $10,000,000	9.2
Over $10,000,000	7.9

On jobs over $100,000, the number of bidders increases as the job size increases, up to about $10,000,000; then the number of bidders tapers off. These jobs cover a wide range of project types, distributed geographically throughout the United States.

An analysis of this type—relating the number of bidders to the size of job (or other important job characteristics)—is useful to the contractor in developing an appropriate bidding strategy for situations in which the exact number and identity of competitors is not known. By anticipating just the approximate amount and type of competition to be encountered, many jobs can be eliminated from further consideration at a great saving in cost, and with no sacrifice of potential profits.

The contractor should always try to predict, by any available means, the number (and identity) of competitors likely to be encountered on each prospective job.

The Spread

The spread—the difference between the low bid and the second low bid—is significant for several reasons:

1. The spread indicates, to some extent, the intensity of competition for a job.
2. The spread measures the amount of money left "on the table," and tells how much higher the low bidder could have been and still taken the job.
3. An unusually wide spread is probably indicative of an estimating error on the low bidder's part, especially if the second and higher bids are grouped closely together. The low bidder in such a case might well weigh the economic consequences of taking the job against the alternative of forfeiting his bid bond.

Based on a sample of 60 recent jobs, ranging in size from $23,000 to $123,000,000 and attracting anywhere from 4 to 15 bidders each, several general observations can be made.

As the number of bidders increases, the percentage spread decreases. Figure 47 shows the relationship between the spread and the number of bidders based on this 60-job sample.

On jobs having from 4 to 6 bidders, the average spread was 8 percent; 7 to 9 bidders resulted in a spread averaging 5.8 percent; for 10 to 12 bidders, the spread averaged 3.8 percent; and jobs attracting 13 to 15 bidders had only a 2 percent spread.

An important conclusion that might be inferred from these data is that on most jobs the low bidder generally bids much lower than is necessary. This is not a valid conclusion, however.

Figure 48 shows the percentage of all jobs having a spread equal to or greater than a given amount. For example, approximately 90 percent of all jobs have a spread of 1 percent or more; 77 percent have at least a 2 percent spread; and 42 percent of all jobs have a spread of 5 percent or more.

Since half of all jobs have a spread of 4 percent or more, there is an even chance that the contractor, by increasing his normal bid by 4 percent, will not affect his chance of being low bidder.

The economic significance of this fact can be illustrated on a job having estimated direct costs of $90,000. If the contractor would normally bid the job at $100,000, he could easily calculate the probable results of increasing his bid.

If he were to increase his bid by 1 percent to $101,000, he would still get 91 percent of the jobs on which he would have been low bidder with a bid of $100,000. Instead of making $10,000, he would have an opportunity to make $11,000 91 percent as many times, an average of $10,010.

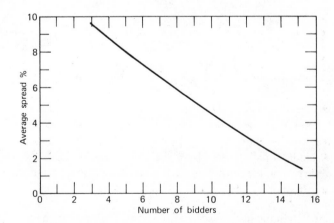

FIGURE 47.   Average spread versus number of bidders.

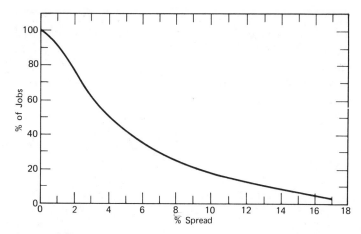

FIGURE 48.   Percentage of all jobs having spread equal to or greater than any given amount.

Raising the bid to $102,000 would yield a gross profit of $12,000 on 77 percent as many jobs as before, an average of $9240. And a bid of $105,000 would yield $15,000 or 42 percent as many jobs as the $100,000 bid, or $6300.

The obvious conclusion to be drawn from these calculations, then, is not to worry about leaving 1 or 2 percent on the table. Trying to decrease the amount in this situation will result only in getting fewer jobs. While an additional 1 percent may be justified, any more would be dangerous.

Again, these inferences apply only to the selected job sample. But a similar analysis, based on the contractor's personal experience, could provide much interesting and useful information.

Distribution of Bids

A common occurrence—one that has amazed outsiders and frustrated many contractors—is the fact that many of the bids submitted on all types of projects appear to be below cost.

During the past few years, nearly 80 percent of all major jobs are let at less than the engineer's estimate; some jobs are even let at less than one-half the engineer's estimate. This peculiar circumstance cannot be entirely attributed to the conservative nature of the engineer's estimates.

According to an anlysis conducted for several general and special trade contractors covering competitors' bids submitted over a recent two-year period, up to 35 percent of all competitors' bids were below the estimated direct out-of-pocket costs of performing the work. Some bids—nearly three percent of the total encountered for one contractor—were placed at more than 30 percent below cost. Admittedly, part of the extreme variances was probably due to this contractor's estimating procedures; still, he was able to hold his own in competition with other reputable firms, and his estimates closely reflected his own costs of operation.

Figures 49 and 50 show how the competitor's bids were distributed for four different types of contractors, expressed as a percentage of these contractors' estimated direct job costs.

The contractors' problems in competitive bidding are vividly illustrated in Figures 49 and 50. To be 100 percent certain of getting a job, he would be

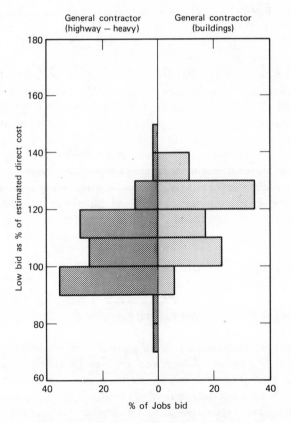

FIGURE 49. Distribution of low bids (general contractor).

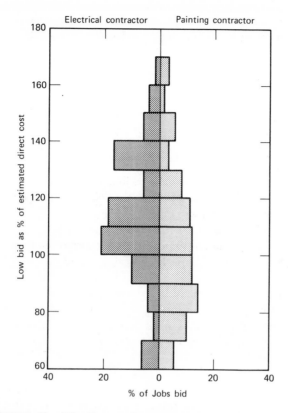

FIGURE 50.   Distribution of low bids (special trade contractors).

forced to bid at about 50 percent of cost; if he were to bid every job at cost, he could still be underbid by about one-third of his competitors; and if he bid all jobs with a 15 percent markup, his bid might beat only one-third of his competitors.

Distribution of Low Bids

Analyzing the distribution of low bids is even more depressing than analyzing the distribution of all bids.

Based on one of the general contractor's experience, the average low bid encountered on 73 jobs ranged from less than one-half the estimated job cost to approximately 25 percent above, with most falling on the low side. The low bids were distributed as follows:

Ratio of Low Bid to Estimated Cost	Number of Jobs	Percentage of Jobs
Less than 0.80	8	11
0.80 to 0.90	13	18
0.90 to 1.00	24	33
1.00 to 1.10	16	22
1.10 to 1.20	10	13
1.20 to 1.30	2	3
Total	73	100

As would be expected, the low bid was found to vary generally according to the number of bidders encountered on the job. As the number of bidders increases, the low bid decreases. Figure 51 shows how the 73 low bids were distributed according to the number of bidders on the job.

Figure 51 also shows the theoretical low bid expected on each job according to the number of bidders. This theoretical low bid curve is based on the contractor's own experience, and represents the bid at which his probability of being low bidder would be exactly 50 percent. In other words, if he were to bid at 92 percent of cost on jobs attracting a total of 16 bids, he would expect to receive about one-half of the jobs at his price. Approximately one-half the low bids fall above the theoretical low-bid line, with the other one-half falling below.

The theoretical low bid on a job having any given number of competitors can be found by identifying the bid required to give a certain probability of underbidding a single typical competitor using the factors in Table 28.

For example, based on the past experience of the contractor in question, he found that a markup of 10 percent would given him about a 50 percent change of getting a job against a single typical competitor. Against two typical competitors, a markup of 4 percent above cost would be required to give him a 50 percent chance of being low bidder; this markup would correspond to a .707 (or 70.7 percent) probability of underbidding a single competitor. Or, stated still another way, if he had a 70.7 percent chance of underbidding one typical competitor, he would have a 50 percent chance of underbidding two.

Similarly, a bid giving 79.4 percent probability of underbidding 1 competitor would be required to give a 50 percent chance of underbidding 3; an 84.1 percent probability of underbidding 1 competitor corresponds to a 50 percent chance of underbidding 4; and to have a 50 percent chance of underbidding 10 competitors at the same time would require that the chance of underbidding each one separately would be 93.3 percent—probably requiring a bid below cost.

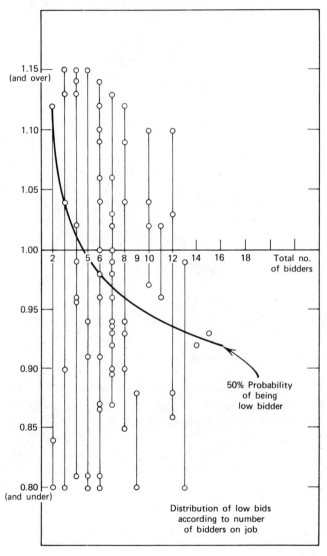

FIGURE 51. Distribution of low bids according to number of bidders on job (general contractor).

TABLE 28. Probability of Underbidding a Single
Competitor Required to Insure 50 Percent Chance of
Underbidding Different Numbers of Competitors

Number of Competitors	Probability of Underbidding One Competitor
1	.500
2	.707
3	.794
4	.841
5	.871
6	.891
7	.906
8	.917
9	.926
10	.933
11	.939
12	.944
13	.948
14	.952
15	.955

Causes of Variations in Bids

Some of the major causes of estimating errors were discussed in Chapter 5; these included the omission of items; undermeasurement of quantities; underestimation of labor requirements; and a failure to recognize all elements of cost, such as overheads.

A cost estimate is never 100 percent accurate. The actual job costs will, instead, be distributed about the estimated cost. Estimating accuracy can be defined by the degree to which actual costs vary from the estimated costs, as well as by the relationships between average estimated costs and average actual costs.

A good detailed estimate should generally be accurate within 5 percent. Even so, on the average, actual costs may vary by as much as 20 percent from the estimated costs due to unforeseen job conditions. If the possibility of unusual job conditions is recognized by some contractors and not by others, those who failed to investigate the job thoroughly are apt to be the low bidders.

Even the best estimators will make outright mistakes, however, that may cause an estimate to be low by as much as 15 to 25 percent. And even if the chance of such an error is only 1 in 100, when 12 different firms compete for the same job, the chances of errors are greatly magnified. In fact, if each of 12 estimators has a 99 percent chance of being accurate within 5 percent, there will still be an estimate that falls outside the 5 percent range on more than 10 percent of the jobs. And few estimators can count on being within 5 percent on 99 jobs in 100.

Table 29 shows how the chances of error increase with the number of bidders. For example, if there were a 95 percent chance that each estimator would be within 5 percent of the actual cost on a given job, there would be only a 60 percent chance that all of 10 estimates would fall within that range. If each estimator was accurate within 5 percent only 80 percent of the time, there would be only an 11 percent chance that all 10 of the estimates would be that accurate; in other words, on almost 9 out of 10 jobs at least one of the estimates would fall outside the 5 percent range.

The situation described by Table 29 is sometimes known as Murphy's Law. Simply stated, Murphy's Law says "the more chances there are for something to go wrong, the greater the chance of it going wrong." Or even simpler, "if something can go wrong, it will."

Seasonal considerations may cause further fluctuations in bid levels. For some reason, many contractors feel that after they have reached a certain volume, "everything is profit from now on." This sometimes results in a complete disregard for overhead and equipment costs, thereby lowering bids to a dangerously low level.

And frequently this situation will lead to still another: desperation. The feeling that any volume, regardless of profit, is better than no volume at all is by no means uncommon. Once a financial position has been reached where money is urgently needed this month to pay last month's bills, money from any source is acceptable—even if the source is this month's jobs taken at cost or below. By

TABLE 29. Probability of Actual Costs Falling within Any Prescribed Range

Number of Bidders	Percentage of Time That Actual Cost Will Fall Within a Specified Percentage of Estimated Cost				
1	95	90	85	80	75
2	90	81	72	64	56
3	86	73	61	51	42
4	81	66	52	41	32
5	77	59	44	33	24
6	74	53	38	26	18
7	70	48	32	21	13
8	66	43	27	17	10
9	63	39	23	13	8
10	60	35	20	11	6

keeping enough jobs coming in, even at a direct out-of-pocket loss, a contractor can stay in business long enough to disrupt completely the profitability of an entire industry.

Summary

The analysis of past bids can provide a wealth of valuable information to the contractor in clarifying the relationships among many different competitive factors. The most meaningful information should be found in the contractor's own files, consisting of job descriptions, estimated costs, and competitors' bids; but published sources can also provide useful information. Relationships having special significance to the contractor include those between the number of bidders and the job characteristics; the spread distribution and the effect of the number of bidders on the spread; the distribution of all bids with respect to estimated job costs; and the distribution of low bids according to the number of bidders. Wide variations in bids are to be expected, with the possibility of estimating errors increasing rapidly as the number of bidders on a job increases.

17

Bidding Strategy in Theory

To despise theory is to have the excessively vain pre-
tention to do without knowing what one does, and to
speak without knowing what one says.

FONTENELLE

The contractor's problem is clear. He can always assure himself of getting a job simply by bidding low enough. But if he does, he probably has little, if any, chance of making a profit. Any contractor can assure himself of making a profit if he does get a job simply by bidding high enough. But if he does, he probably has little, if any, chance of getting a job.

The solution is clear: Bid high encough to make a profit, and low enough to get a job—both at the same time. This is the objective of a bidding strategy.

There are several obvious, but nevertheless important, points that should be emphasized in developing an appropriate bidding strategy:

1. Being the low bidder is not always desirable; low bidders are usually the first to go bankrupt.
2. Low bidders are generally followed into bankruptcy by high bidders; trying for too high a profit may result in getting none.
3. The only way to make a profit is to bid every job at a profit.
4. There is little merit in getting volume just for the sake of volume. The object of being in business is to make a profit, and no amount of volume can take the place of even a moderate profit.

Theories of Competitive Bidding

Many different theoretical approaches to competitive bidding have been proposed and tested with varying results. Any of these strategies should

211

improve the contractor's bidding effectiveness, and whichever one works best for a particular competitive situation is obviously the best one to use. It will be well worth whatever time is required to at least become familiar with the different approaches; they all offer some good ideas, and "even a bad plan is better than no plan at all."

Probably the two best-known and widely accepted theoretical approaches are known as (1) Friedman's Model and (2) Gates' Model. The approach described in this chapter is based on Friedman's Model. While this model has been disputed for some time, it has the major virtue of simplicity, offering a straightforward graphical approach to bidding strategy; it has proven effective for many contractors.

The main point of contention with Friedman's Model has to do with whether the competitive bids submitted on a given project are actually independent of each other, an argument that certainly merits consideration. Obviously, all contractors bidding a certain job are influenced to some extent by the specific conditions—both physical and competitive—expected to be encountered on the job.

Other approaches attempt to take these common conditions into account by means of more sophisticated and complex mathematical techniques, adding sometimes complicated features intended to improve the results. Whether these refinements yield any measureable improvements over the more straightforward Friedman approach is questionable. Research can "prove" almost anything, but increased profits are the only true test of any bidding strategy.

The Databid System, presented in Chapter 19, offers a nontheoretical approach to developing a competitive bidding strategy based entirely on data accumulated from different jobs of varying characteristics. The Databid System takes into account the total competitive situation without worrying about any mathematical relationships or theories.

Sources of additional information on the various approaches to competitive bidding are cited in the Appendix.

Basic Concepts

Disregarding errors, the lower limit of bids is generally set by the estimated direct cost of a given project. The relationship between the bid price and the estimated cost depends on several factors, such as the contractor's need for work, the minimum acceptable markup, and the maximum he thinks he can get.

Every contractor realizes that his chances of being low bidder have a direct relationship to his bid. The higher his bid, the lower his chances of its being successful; the lower the bid, the better his chances of getting a job. In extreme cases the contractor could do either of the following:

1. Bid so low that he is sure of getting the job, even though there would be no profit.
2. Bid so high that a large profit is assured, even though his chances of getting the job would be nil.

The basic concept underlying the competitive bidding strategy consists simply in recognizing that there is some one bid which results in the best possible combination of two factors:

1. The profit resulting from obtaining a contract at a specified bid price.
2. The probability of getting the job by bidding that amount.

Bidding Against a Single Competitor

Figure 52 illustrates the effect of the bid price on the chances of being low bidder, when bidding against a single competitor. In Figure 52 the contractor can be certain of being low bidder only if he bids the job at cost. By bidding at 10 percent above cost, he can expect to be low bidder on 60 percent of the jobs; a 20 percent markup will be low on 20 percent of the jobs; and a 25 percent markup will result in no jobs at all.

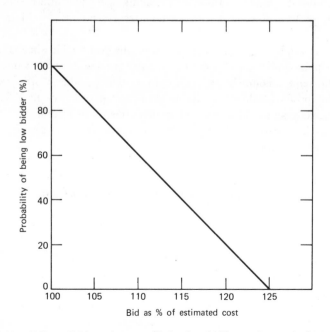

FIGURE 52. Effect of bid on chances of being low bidder against a single competitor.

A few of the possible alternatives and their results are as follows:

1. Bid at cost and get all the jobs but no profits.
2. Bid 5 percent above cost, and get 80 percent of the jobs bid. The average profit per job bid—the "expected" profit—would then be 80 percent of 5 percent, or 4 percent.
3. Bid 10 percent above cost, and get 60 percent of the jobs bid, for an expected profit of 6 percent per job bid.
4. Bid 12.5 percent above cost, and get half the jobs, for an expected profit of 6.25 percent.
5. Bid 15 percent above cost, and get 40 percent of the jobs, with an expected profit of 6 percent.
6. Bid 20 percent above cost, and get 20 percent of the jobs, giving an expected profit of 4 percent.
7. Bid 25 percent above cost and get no jobs and no profit.

Figure 53 shows the expected profit associated with each combination of markup and probability of being low bidder, when bidding against this one competitor. In this example the contractor will, in the long run, make more money by bidding at 12.5 percent above cost than he could make either by taking more jobs at a lower markup, or by taking fewer jobs at a higher markup. A markup of 12.5 percent above cost, then, represents his "optimum" or best possible bid.

In this example the expected profit resulting from a bid that is above the optimum point—above 12.5 percent, in this case—will be the same as a bid placed the same amount below the optimum bid. In other words, exactly the same total profit would result from getting 60 percent of the jobs at a 10 percent markup as would be realized from getting 40 percent of the jobs at a 15 percent markup. But the higher markup on fewer jobs would be far less risky and far more desirable.

Bidding Against More Than One Competitor

The preceding example assumed that only one competitor was involved in the bidding, and that the probability of underbidding this one competitor varied inversely with the bid price as shown in Figure 52. If several similar competitors are involved in the bidding, however, the picture changes considerably.

In Chapter 13, the fourth rule of probability stated that "the probability that two or more independent events will happen simultaneously is equal to their individual probabilities all multiplied together." In bidding, *all* competitors

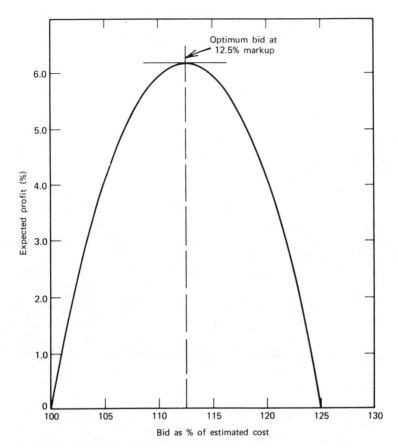

FIGURE 53. Effect of bid on expected profit against a single competitor.

must be underbid at the same time; therefore, the probability of underbidding several competitors at the same time is equal to the probabilities of underbidding each one separately, all multiplied together, *providing* that their individual bids can be considered totally independent of each other.

Figure 54 illustrates how the probability of being low bidder varies according to the number of competitors involved in the bidding, assuming the probability of underbidding a single competitor is still as indicated in Figure 52.

In Figure 54 the top line is the same as in Figure 52, representing the probability of underbidding a single competitor; this line indicates that a bid of 112.5 percent of cost—a 12.5 percent markup—will underbid the single competitor 50 percent of the time.

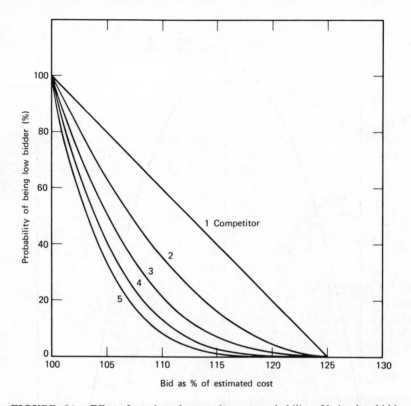

FIGURE 54. Effect of number of competitors on probability of being low bidder.

If two competitors are involved, the probability of underbidding both at the same time will be only 50 percent of 50 percent, or 25 percent. Similarly, the chances of underbidding three competitors will again be reduced by half—.50 × .50 × .50—to 12.5 percent. Four competitors gives only a 6.25 percent chance of being low bidder with a 12.5 percent markup, and five competitors reduces the chances to only 3.125 percent.

The other points on the curves in Figure 54 are calculated in the same way. A 5 percent markup, for example, gives an 80 percent chance against one competitor, a 64 percent against two (.80 × .80), a 51.2 percent chance against three (.80 × .80 × .80), a 40.96 percent chance against four (.80 × .80 × .80 × .80), and a 32.768 percent chance against five competitors (.80 × .80 × .80 × .80 × .80).

Figure 55 shows the effect of these decreased probabilities of being low bidder on the job's profit expectations. With a 12.5 percent markup, the contractor could get half the jobs against a single competitor, for an expected profit of 6.25 percent. Against two competitors, however, he can expect to get only 25 percent of the jobs at this markup, reducing his profit expectation to 25 percent of 12.5 percent, or only 3.125 percent.

Similar effects are felt at each markup. As the number of bidders increases, the probability of getting the job with any given bid decreases, and the expected profit associated with any specified markup also decreases. Also, the bid which will result in the highest possible long-run profits—the optimum bid—will decrease as the number of bidders increases. Figure 55 indicates that, in this example, the optimum bid against one competitor is 12.5 percent, with an expected profit of 6.25 percent. Against two competitors, the optimum bid drops to about 8 percent, with an expected profit of 3.8 percent. Against three competitors a markup of 7 percent will yield the highest possible expected profit, 2.6 percent. Against four competitors the optimum bid is 5 percent, giv-

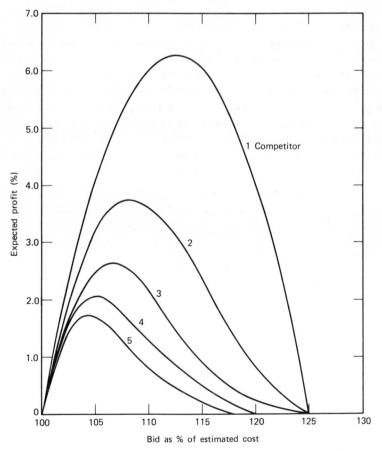

FIGURE 55. Effect of number of competitors on expected profits.

ing an expected profit of 2.1 percent. And against five competitors a markup of 4 percent will result in the maximum expected profit of 1.7 percent.

Determining the Probability of Being Low Bidder

One of the key steps in developing a successful competitive bidding strategy consists of identifying the probabilities of underbidding competitors—or, in other words, in determining the "odds of the game."

The probability curve of Figure 52 was a straight line, ranging from a 100 percent chance of being low bidder with a bid placed at cost to a 0 percent chance of getting the job at a 25 percent markup. Although this type of probability distribution is entirely possible, few competitors are likely to follow so regular a pattern. Most competitors' bidding patterns will more nearly resemble the distribution shown in Figure 56.

Figure 56 represents a "normal" distribution of a competitor's bids, as related to the contractor's own estimated direct costs of performing the work. In this example, the competitor was found to bid at between 100 and 105 percent of estimated direct cost on 5 percent of the jobs; between 105 and 110 percent of cost on 25 percent of the jobs; between 110 and 115 percent of cost on 40 percent of the jobs; between 115 and 120 percent of cost on 25 percent of the jobs; and from 120 to 125 percent of cost on 5 percent of the jobs.

The bidding distribution pattern of this competitor could have been found simply by tabulating his bids on all jobs for which cost estimates were made, in

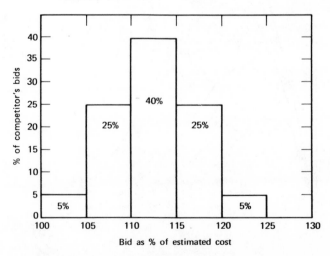

FIGURE 56.   Normal distribution of a competitor's bids.

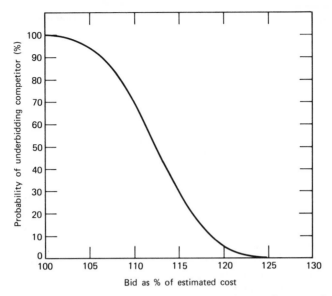

FIGURE 57. Probability curve based on normal distribution of a competitor's bids.

each case relating the competitor's bids to the estimated job costs. Wide variations will be found in the bidding characteristics exhibited by different competitors; competitors' bids may range from less than half to more than double the estimated job cost, with the extreme variations most likely caused by errors or oversights on the low side, and by a complete lack of interest in getting the job on the high side (perhaps a "complimentary" bid).

From the frequency distribution of the competitor's past bids, a probability curve can be constructed, giving the chances of underbidding this competitor with any given bid. Figure 57 shows how the probability curve looks for the frequency distribution of Figure 56.

Figure 57 indicates that a bid placed at cost will result in underbidding this competitor on every job; in other words, on 100 percent of all past jobs involving this competitor, this competitor bid above the estimated job cost.

If a bid were placed at 5 percent above cost, the bid would have been lower than 95 percent of the competitor's bids on past jobs, for only 5 percent of his bids fell within the 100 to 105 percent of cost range. Similarly, a markup of 10 percent would have been higher than 30 percent of this competitor's bids—the 5 percent falling in the 100 to 105 percent range and the 25 percent falling in the 105 to 110 percent—and lower than the remaining 70 percent of this competitor's bids. Thus the probability of underbidding this competitor by bidding at 10 percent above cost would be 70 percent. The remaining points on the probability curve of Figure 57 are determined the same way, with each point defining the percentage of the competitor's bids which experience has shown to fall above that level.

Determining the Expected Profit

A profit expectation curve can easily be developed from the probability curve. The expectation curve, shown in Figure 58, gives the average long-run profit resulting from any given level of markup when bidding against a competitor having these bidding characteristics.

Figure 58 indicates that a markup of about 10 percent against this competitor would have yielded the highest possible profits; such a markup would have captured 70 percent of the jobs, giving an average profit per job bid of 70 percent of 10 percent, or 7 percent. A lower bid would have captured more jobs, and a higher bid would have resulted in higher profits on the jobs won; but neither a higher nor a lower bid would have equaled the total profit associated with a 10 percent markup in this case. A 5 percent markup would have taken 95 percent of the jobs, but the profit would have been only 4.75 percent per job bid; a 15 percent markup would have taken 30 percent of the jobs, for an average 4.5 percent profit.

Probability of Underbidding Combinations of Competitors

Every competitor will exhibit different bidding characteristics; some bid consistently high, some bid consistently low, some spread their bids uniformly

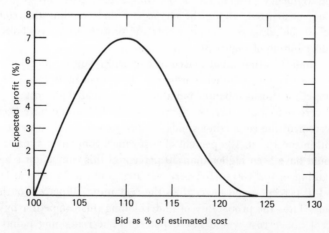

FIGURE 58. Expected profit curve based on normal distribution of a competitor's bids.

over a wide range, and some may bid within fairly well defined and narrow limits. The strategy to be employed against each must therefore vary to take maximum advantage of each one's individual characteristics and weaknesses.

Figure 59 shows three different bidding distribution patterns representing three different competitors. Competitor A (Figure 59a) has the same characteristics as used in the preceding example. Competitor B (Figure 59b) tends to group his bids lower than competitor A, with 40 percent of his bids falling in the 105 to 110 percent range. Competitor C (Figure 59c) generally bids higher than the other two, with 40 percent of his bids grouped in the 115 to 120 percent of cost range.

When bidding against these three competitors, a given markup will offer different chances of underbidding each one. For example, a 10 percent markup gives a 70 percent chance of underbidding competitor A, a 45 percent chance of underbidding competitor B, and an 85 percent chance of underbidding competitor C. And since the probability of underbidding combinations of competitors is equal to their individual probabilities multiplied together, the chances of getting a job at any given markup are sharply reduced when more than one competitor is involved. Figure 60 shows the probability of underbidding different combinations of competitors A, B, and C.

Figure 60a assumes competition against A and B. Since the probability of underbidding A with a 10 percent markup is 70 percent, and the corresponding probability of underbidding competitor B with the same markup is 45 percent, the probability of underbidding both A and B at the same time is .70 × .45, or only 31.5 percent. The probability of underbidding both A and B with any specified bid can be found by identifying the probability of underbidding each one separately, then multiplying these probabilities together.

Figures 60b, 60c, and 60d shows the probabilities associated with other combinations of competitors. For example, a 10 percent markup gives a 31.5 percent chance of underbidding A and B; a 59.5 percent chance of underbidding A and C (.70 × .85); a 38.25 percent chance of underbidding B and C (.45 × .85); and a 26.775 percent chance of underbidding all three competitors (.70 × .45 × .85).

Profit Expectation Against Combinations of Competitors

Again, as the probability of being low bidder varies according to the amount and type of competition, the expected profit associated with each job also varies. Figure 61 shows the profit expectation curves associated with each of the four probability curves of Figure 60.

When bidding against competitors A and B, the optimum bid is found to occur at a markup of about 7 percent; this markup allows a 64 percent chance of getting the job for an expected profit of some 4.5 percent per job bid.

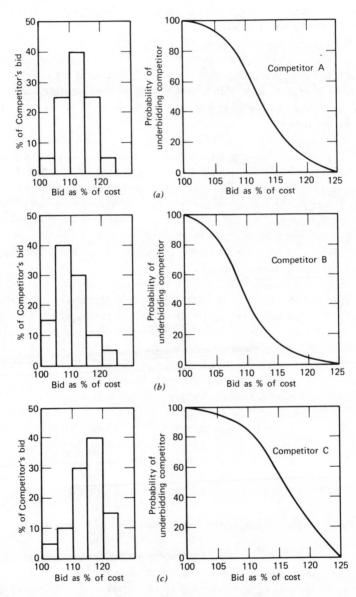

FIGURE 59. Bid distribution and probability curves for different competitors.

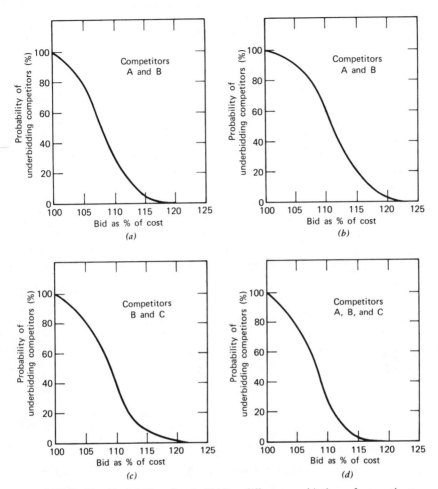

FIGURE 60.   Probability of underbidding different combiations of competitors.

Against A and C, the optimum bid is higher (because C generally bids higher than did B), occurring at a markup of about 10 percent, at which point there is a 60 percent chance of being the low bidder, and an expected profit of 6 percent.

Against B and C—the lowest and the highest bidders—the optimum bid is 7 percent above cost, with a corresponding probability of 68 percent and a maximum expected profit of 4.76 percent. Note that the optimum bid against B and C is the same as when bidding against A and B, but that the expected profit is slightly higher.

Finally, against all three competitors, the optimum bid is also at about a 7 percent markup, where a 60 percent chance of being low will yield an expected

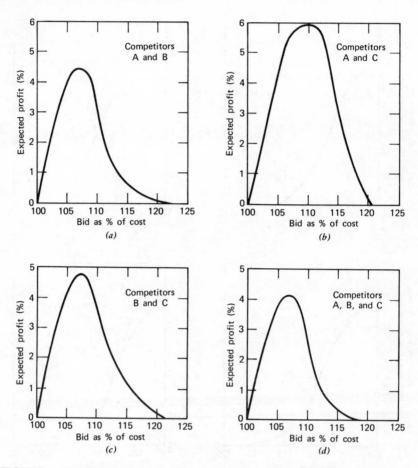

FIGURE   61. Expected profit when bidding against different combinations of competitors.

profit of 4.2 percent. Again, the optimum bid in this case is the same as in two of the other cases—a 7 percent markup—but the expected profit is still lower, because of the increased competition.

Each of these four situations, then, involves different profit expectations. If a choice were offered, the contractor would do well to bid against A and C first, followed by B and C, then A and B, and last, all three.

The Typical Competitor

When individual competitors and their bidding characteristics can be identified in advance, the best results can usually be obtained by considering them individually as in the previous examples. However, there are apt to be relatively few jobs on which all competitors can be identified, or where sufficient data are available to determine properly the bidding characteristics of all participants. In such cases the concept of the "typical"—or average—competitor can be used to advantage.

The typical competitor is simply a composite made up of all bids of all competitors. As such the typical competitor refers to no one competitor in particular, but to all competitors in general. Figure 62 shows the bid frequency

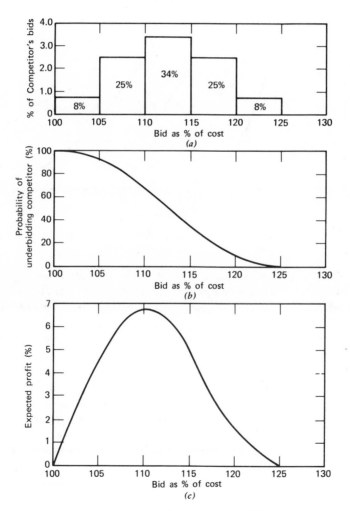

FIGURE 62. Characteristics of a typical competitor. (*a*) Distribution of competitor's bids; (*b*) probability curve; (*c*) expected profit curve.

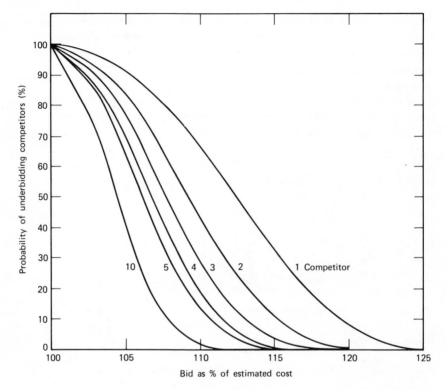

FIGURE 63. Probability of underbidding different numbers of typical competitors.

distribution, the probability curve, and the profit expectation curve made up of the three individual competitors having the characteristics described in Figure 59. By adding together the individual characteristics, the curves of Figure 62 were developed.

Figure 62*a* shows the average distribution of all bids encountered from the three competitors; 8 percent were in the 100 to 105 percent of cost range, 25 percent fell between 105 and 110 percent of cost, 34 percent were in the 110 to 115 percent range, 25 percent were in the 115 to 120 percent range, and the remaining 8 percent were between 120 and 125 percent of estimated cost.

This frequency distribution results in the probability curve shown in Figure 62*b*. A markup of 5 percent against a single typical competitor will win 92 percent of the jobs; a 10 percent markup will be low on 67 percent of the jobs; a 15 percent markup will be low bid on 33 percent of the jobs; a 20 percent markup will win 8 percent of the jobs; and a 25 percent markup will result in getting none of the jobs.

These probabilities result in the profit expectations shown in Figure 62*c*. A 92 percent chance of making a 5 percent profit gives an expected profit of 4.6

percent; a 67 percent probability of a 10 percent profit averages 6.7 percent; a 33 percent chance of a 15 percent profit yields an average of 4.95 percent; and an 8 percent chance of a 20 percent profit results in an expected profit of 1.6 percent. This optimum bid against this one average competitor, then, is at a markup of 10 percent; this point represents the best possible combination of the markup and the chances of getting the job. A lower bid will result in less profit because of the lower profit on each job; a higher bid will result in less profit because of the fewer jobs obtained.

The concept of a typical competitor is especially valuable when bidding against numerous unknown competitors. By using this concept, the general level of bids likely to result in maximum profits can be identified and used as a guide in setting an exact price or in identifying the most potentially profitable jobs.

Bidding Against Different Numbers of Typical Competitors

The probability of being low bidder when bidding against different numbers of typical competitors can be caluclated in the same manner as when the specific competitors are known. The probability of underbidding any given number of typical competitors can be found by first taking the probability of underbidding a single typical competitor, then multiplying this number by itself that same number of times. In other words, the probability (P) of underbidding n typical competitors is

$$P = C^n$$

where C is the probability of underbidding one competitor with a specific markup. Table 30 is presented as a convenience in making these calculations and gives the probability of underbidding up to 10 typical competitors based on the probability of underbidding each one separately.

Referring to Table 30, we see that if the probability of underbidding a single typical competitor were .85, then the probability of underbidding 5 typical competitors would be .444, or 44.4 percent; the probability of underbidding 10 competitors would be .197, or 19.7 percent.

Again using the typical competitor identified in the preceding section—a combination of Competitors A, B, and C from Figure 59—the probability of underbidding different numbers of competitors can be easily determined. Table 31 shows the probability of underbidding different numbers of typical competitors having the bidding characteristics described by Figure 62.

The probability of underbidding one competitor is the same as shown in Figure 28B. The probability of underbidding any number of typical competitors is found by multiplying the probability of underbidding a single typical

TABLE 30. Probability of Underbidding Various Numbers of Competitors

				Number of Competitors					
1	2	3	4	5	6	7	8	9	10
1.000	1.000	1.000	1.000	1.000	1.000	1.000	1.000	1.000	1.000
.990	.980	.970	.961	.951	.941	.932	.923	.914	.904
.980	.960	.941	.922	.904	.886	.868	.851	.834	.817
.970	.941	.913	.885	.859	.833	.808	.784	.760	.737
.960	.922	.885	.849	.815	.783	.751	.721	.693	.665
.950	.903	.857	.815	.774	.735	.698	.663	.630	.599
.940	.884	.831	.781	.734	.690	.648	.610	.573	.539
.930	.865	.804	.748	.696	.647	.602	.560	.520	.484
.920	.846	.779	.716	.659	.606	.558	.513	.472	.434
.910	.828	.754	.686	.624	.568	.517	.470	.426	.389
.900	.810	.729	.656	.590	.531	.478	.430	.387	.349
.890	.792	.705	.627	.558	.497	.442	.394	.350	.312
.880	.774	.681	.600	.528	.464	.409	.360	.316	.279
.870	.757	.659	.573	.498	.434	.377	.328	.286	.248
.860	.740	.636	.547	.470	.405	.348	.299	.257	.221
.850	.723	.614	.522	.444	.377	.321	.272	.232	.197
.840	.706	.593	.498	.418	.351	.295	.248	.208	.175
.830	.689	.572	.475	.394	.327	.271	.225	.187	.155
.820	.672	.551	.452	.371	.304	.249	.204	.168	.137
.810	.656	.531	.430	.349	.282	.229	.185	.150	.122
.800	.640	.512	.410	.328	.262	.210	.168	.134	.107
.790	.624	.493	.390	.308	.243	.192	.152	.120	.095
.780	.608	.475	.370	.289	.225	.176	.137	.107	.085
.770	.593	.457	.352	.271	.208	.160	.124	.095	.073
.760	.578	.439	.334	.254	.193	.146	.111	.085	.064
.750	.563	.422	.316	.237	.178	.133	.100	.075	.056
.740	.548	.405	.300	.222	.164	.122	.090	.067	.049
.730	.533	.389	.284	.207	.151	.110	.081	.059	.043
.720	.518	.373	.269	.193	.139	.100	.072	.052	.037
.710	.504	.358	.254	.180	.128	.091	.065	.046	.033
.700	.490	.343	.240	.168	.118	.082	.058	.040	.028
.690	.476	.329	.227	.156	.108	.074	.051	.035	.024
.680	.462	.314	.214	.145	.099	.067	.046	.031	.021
.670	.449	.301	.202	.135	.090	.061	.041	.027	.018
.660	.436	.287	.190	.125	.083	.055	.036	.024	.016
.650	.423	.275	.179	.116	.075	.049	.032	.021	.013
.640	.410	.262	.168	.107	.069	.044	.028	.018	.012
.630	.397	.250	.158	.099	.063	.039	.025	.016	.010
.620	.384	.238	.148	.092	.057	.035	.022	.014	.008

TABLE 30. *(Continued)*

				Number of Competitors					
1	2	3	4	5	6	7	8	9	10
.610	.372	.227	.138	.084	.052	.031	.019	.012	.007
.600	.360	.216	.130	.078	.047	.028	.017	.010	.006
.590	.348	.205	.121	.071	.042	.025	.015	.009	.005
.580	.336	.195	.113	.066	.038	.022	.013	.007	.004
.570	.325	.185	.106	.060	.034	.020	.011	.006	.004
.560	.314	.176	.098	.055	.031	.017	.010	.005	.003
.550	.303	.166	.092	.050	.028	.015	.008	.005	.003
.540	.292	.157	.085	.046	.025	.013	.007	.004	.002
.530	.281	.149	.079	.042	.022	.012	.006	.003	.002
.520	.270	.141	.073	.038	.020	.010	.005	.003	.001
.510	.260	.133	.068	.035	.018	.009	.005	.002	.001
.500	.250	.125	.063	.031	.015	.008	.004	.002	.001
.490	.240	.118	.058	.028	.014	.007	.003	.002	.001
.480	.230	.111	.053	.025	.012	.006	.003	.001	.001
.470	.221	.104	.049	.023	.011	.005	.002	.001	.001
.460	.212	.097	.045	.021	.009	.005	.002	.001	.000
.450	.203	.091	.041	.018	.008	.004	.002	.001	.000
.440	.194	.085	.037	.016	.007	.003	.001	.001	.000
.430	.185	.080	.034	.015	.006	.003	.001	.001	.000
.420	.176	.074	.031	.013	.005	.002	.001	.000	
.410	.168	.069	.028	.012	.005	.002	.001	.000	
.400	.160	.064	.026	.010	.004	.002	.001	.000	
.390	.152	.059	.023	.009	.004	.001	.001	.000	
.380	.144	.055	.021	.008	.003	.001	.000		
.370	.137	.051	.019	.007	.003	.001	.000		
.360	.130	.047	.017	.006	.002	.001	.000		
.350	.123	.043	.015	.005	.002	.001	.000		
.340	.116	.039	.014	.005	.002	.001	.000		
.330	.109	.036	.012	.004	.001	.000			
.320	.102	.033	.010	.003	.001	.000			
.310	.096	.030	.009	.003	.001	.000			
.300	.090	.027	.008	.002	.001	.000			
.290	.084	.024	.007	.002	.001	.000			
.280	.078	.022	.006	.002	.000				
.270	.073	.020	.005	.001	.000				
.260	.068	.018	.005	.001	.000				
.250	.063	.016	.004	.001	.000				
.240	.058	.014	.003	.001	.000				
.230	.053	.012	.003	.001	.000				

TABLE 30.   *(Continued)*

				Number of Competitors					
1	2	3	4	5	6	7	8	9	10
.220	.048	.011	.002	.001	.000				
.210	.044	.009	.002	.000					
.200	.040	.008	.002	.000					
.190	.036	.007	.001	.000					
.180	.032	.006	.001	.000					
.170	.029	.005	.001	.000					
.160	.026	.004	.001	.000					
.150	.023	.003	.001	.000					
.140	.020	.003	.000						
.130	.017	.002	.000						
.120	.014	.002	.000						
.110	.012	.001	.000						
.100	.010	.001	.000						
.090	.008	.001	.000						
.080	.006	.001	.000						
.070	.005	.000							
.060	.004	.000							
.050	.003	.000							
.040	.002	.000							
.030	.001	.000							
.020	.000								
.010	.000								

competitor by itself that number of times. For example, the probability of underbidding one typical competitor with a markup of 11 percent is 60 percent, or .60; the probability of underbidding 2 competitors, then, is .60 × .60, or 36 percent; of underbidding 3 competitors, $.60^3$, or 21.6 percent; 4 competitors, $.60^4$, or 13 percent; 5 competitors, $.60^5$, or 7.8 percent; and 10 competitors, $.60^{10}$, or 0.6 percent. (These probabilities could also be looked up directly in Table 30.) The effect of the number of competitors on the probability of being low bidder is shown graphically in Figure 63.

Different profit expectations are also associated with different numbers of competitors. These are summarized in Table 32.

Against one typical competitor the expected profit begins at zero where the bid is at cost, increasing to 6.7 percent at a bid of 10 percent above cost, then decreases back to zero when the markup reaches 25 percent.

Against more than one competitor, the expected profit increases to some lesser value, reaching a maximum at a lower markup than is the case against a single competitor. Two typical competitors yield a maximum expected profit of 4.99 percent at an 8 percent markup; with three competitors, a maximum expected profit of 4.15 percent is possible, at a 7 percent markup; against four competitors, the optimum bid is at 6 percent above cost; and against five competitors, the optimum markup is 5 percent. Ten competitors reduces the optimum bid to 4 percent above cost, and results in only a 2.4 percent expected profit. The expected profit curves for different numbers of typical competitors are plotted in Figure 64.

An interesting point that is evident from Figure 64 is that a bid above the optimum value will result in a higher expected profit than a bid that is the same amount below the optimum, regardless of the number of bidders involved

TABLE 31.   Example of Probability Calculations for Typical Competitors

Bid as Percentage of Cost	Probability of Underbidding Different Numbers of Competitors (%)					
	1	2	3	4	5	10
100	100.0	100.0	100.0	100.0	100.0	100.0
101	99	98.0	97.0	96.1	95.1	90.4
102	98	96.0	94.1	92.2	90.4	81.7
103	97	94.1	91.3	88.5	85.9	73.7
104	95	90.3	85.7	81.5	77.4	59.9
105	92	84.6	77.9	71.6	65.9	43.4
106	88	77.4	68.1	60.0	52.8	27.9
107	84	70.6	59.3	49.8	41.8	17.5
108	79	62.4	49.3	39.0	30.6	9.5
109	73	53.3	38.9	28.4	20.7	4.3
110	67	44.9	30.1	20.2	13.5	1.8
111	60	36.0	21.6	13.0	7.8	0.6
112	53	28.1	14.9	7.9	4.2	0.2
113	47	22.1	10.4	4.9	2.3	0.1
114	40	16.0	6.4	2.6	1.0	0.0
115	33	10.9	3.6	1.2	0.4	0.0
116	27	7.3	2.0	0.5	0.1	0.0
117	21	4.4	0.9	0.2	0.0	
118	16	2.6	0.4	0.1	0.0	
119	12	1.4	0.2	0.0		
120	8	0.6	0.1	0.0		
121	5	0.3	0.0			
122	3	0.1	0.0			
123	2	0.0				
124	1	0.0				
125	0	0.0				

TABLE 32.   Example of Calculation of Profit Expectation Against Typical Competitors

Bid as Percentage of Cost	Expected Profit When Bidding Against Different Numbers of Competitors (%)					
	1	2	3	4	5	10
100	0	0	0	0	0	0
101	0.990	0.980	0.970	0.961	0.951	0.909
102	1.960	1.920	1.882	1.844	1.808	1.634
103	2.910	2.823	2.739	2.655	2.577	2.214
104	3.800	3.612	3.428	3.260	3.096	2.396
105	4.600	4.235	3.895	3.585	3.300	2.175
106	5.280	4.644	4.092	3.600	3.168	1.668
107	5.880	4.942	4.151	3.486	2.926	1.225
108	6.320	4.992	3.944	3.120	2.456	0.760
109	6.570	4.797	3.501	2.556	1.863	0.387
110	6.700	4.490	3.000	2.020	1.350	0.180
111	6.600	3.960	2.376	1.419	0.858	0.066
112	6.360	3.372	1.788	0.948	0.504	0.024
113	6.110	2.873	1.352	0.637	0.299	0.013
114	5.600	2.240	0.896	0.364	0.140	0.000
115	4.950	1.635	0.540	0.180	0.060	
116	4.320	1.168	0.320	0.080	0.016	
117	3.570	0.748	0.153	0.034	0.000	
118	2.880	0.468	0.072	0.018		
119	2.280	0.266	0.038	0.000		
120	1.600	0.072	0.020			
121	1.050	0.066	0.000			
122	0.660	0.022				
123	0.460	0.000				
124	0.240					
125	0.000					

in the competition. When some doubt exists regarding the best bid on a particular job, the natural tendency is to bid lower; more profits could generally be achieved in the long run, however, by bidding slightly higher.

Figure 65 shows the relationship between the optimum bid and the expected profit for different numbers of typical competitors. As the number of bidders increases, both the optimum bid and the expected profit decrease. As the number of bidders increases, the expected profit more closely approaches zero—meaning that the low bids are approaching the direct cost of performing the work.

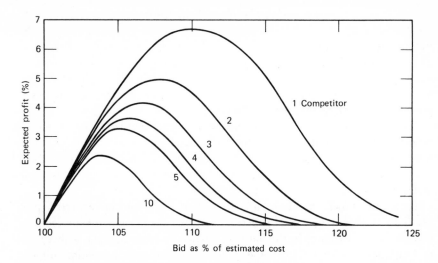

FIGURE 64. Expected profit curves for different numbers of typical competitors.

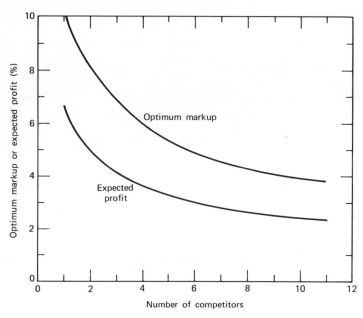

FIGURE 65. Effect of number of competitors on optimum bid and expected profit.

233

Choosing the Best Opportunities

Frequently the contractor is offered a choice among several opportunities. Because of time, financial, or personnel limitations, he may be unable to pursue all of the available opportunities. In such a situation he should therefore apply his resources to only those alternatives that offer the greatest possibilities for profit as measured by their expected profits.

Sometimes several jobs will appear at the same time that offer potentially good profit opportunities but at a time when perhaps the contractor is for some reason unable to place each bid at the optimum level. Instead he must spread some maximum total bid among several jobs so as to maximize his profits under the prescribed conditions or limitations.

Table 33 shows the necessary data. For each job the expected profit associated with any given bid is determined, and the change in expected profit resulting from a change in the bid is calculated. In this example, if there were no limitations on the total bid, the contractor would enter his optimum bid on each job. If, however, the total of his three optimum bids was greater than the total amount allowed for the three, he must decide which prices can be lowered without lowering his expected profit any more than is absolutely necessary. This type of situation might be encountered when several jobs are to be conducted for the same owner, who has only a fixed amount of capital available to cover all the jobs.

TABLE 33.   Determining an Optimum Combination of Bids When Total Bid is Limited ($)

	Job 1			Job 2			Job 3	
Amount Bid	Expected Profit	Change in Expected Profit	Amount Bid	Expected Profit	Change in Expected Profit	Amount Bid	Expected Profit	Change in Expected Profit
100,000	0		100,000	0		100,000	0	
102,000	1,800	+1,800	102,000	1,700	+1,700	102,000	1,900	+1,900
104,000	3,400	+1,600	104,000	3,500	+1,800	104,000	3,800	+1,900
106,000	4,400	+1,000	106,000	5,300	+1,800	106,000	5,300	+1,500
108,000	4,500	+100	108,000	5,800	+500	108,000	6,300	+1,000
110,000	3,100	−1,400	110,000	6,100	+300	110,000	6,600	+300
112,000	1,500	−1,600	112,000	5,600	−500	112,000	6,800	+200
114,000	900	−600	114,000	3,800	−1,800	114,000	5,600	−1,200
116,000	500	−400	116,000	2,100	−1,700	116,000	4,300	−1,300
118,000	300	−200	118,000	1,000	−1,100	118,000	2,700	−1,600
120,000	100	−100	120,000	100	−900	120,000	1,500	−1,200
122,000	0	−100	122,000	0	−100	122,000	700	−800

If there were no limitations on the bids, then the contractor should bid $108,000 on job 1, $110,000 on job 2, and $112,000 on job 3. His total bid would be $330,000, and his expected profit, $17,400.

If the total of all three bids was limited to $328,000, he would lose less in his total expected profit by reducing the bid on job 1 to $106,000, where profit expectation would decrease by only $100. The total of all three bids would thus be $328,000 with a profit expectation of $17,300.

A total bid limitation of $326,000 would result in a further reduction of expected profit. The smallest reduction would be suffered on job 3, where a bid of $110,000 would decrease expected profits by only $200. The three-bid total would now be $326,000 and the expected profit, $17,100.

The total bid can be reduced in the same way to any prescribed level, always reducing the bid so as to lower the total expected profit as little as possible. A total bid of $324,000 could best be satisfied by bidding $106,000 on job 1, $110,000 on job 2, and $108,000 on job 3. Similarly, $322,000 in bids would be distributed among the three jobs at $106,000, $108,000, and $108,000, respectively. And $320,000 would be divided among them at $106,000, $106,000, and $108,000.

The same procedure can be followed regardless of the number or size of the jobs, whenever restrictions are in effect regarding the total amount of capital to be committed. Starting at the optimum value for each job, reductions would be made in the individual bids, each time reducing the bid that would have the least impact on expected profits until the desired total is reached.

Summary

As a contractor's bid on a job increases, his chances of getting the job decrease; but lowering his bid to improve the chances of being low bidder reduces the amount of profit to be made on the job. By applying a competitive bidding strategy, however, he will be able to maximize his profits by identifying the best possible combination of (1) the profit resulting from obtaining a contract at a specified bid price, and (2) the probability of getting the job by bidding that amount. In developing a competitive bidding strategy, the contractor must analyze the bidding characteristics of his competitors. By identifying the distribution of their bids in relation to his own estimated job costs, he can obtain a frequency distribution that can be used to construct a probability curve, showing his chances of being low bidder by bidding any given amount; from the probability curve, the expected profit associated with each bid can be determined. Different combinations of competitors can be considered, and an optimum bid can be calculated for almost any situation so long as the identity or number of competitors can be established in advance of the bidding. The optimum bid depends both on the number of competitors and on their individual characteristics. As the number of competitors increases, the optimum bid and the expected profit both decrease. By computing the expected profit in different situations, jobs offering the most desirable profit opportunities can be easily identified.

18

Matching the Markup to the Job

Theory is the guide to practice, and practice the
ratification and life of theory.

JOHN WEISS

Competitive bidding is the purest type of competitive activity that can be found; it represents, essentially, what economists refer to as "perfect competition."

Under such a system of near-perfect competition, no one individual or firm can control the price at which a contract is let, since the price will be set by the lowest bidder and will be completely independent of the prices submitted by other competitors.

Since a contractor's bid on any given job exerts such an important influence on his chances of getting the job, a great deal of thought is required in deciding on the exact amount of a bid. Being one percent too low can be tolerated; a bid that is one percent too high, however, might just as well have never been submitted at all.

Nearly all contractors employ a bidding strategy of some type. Usually the strategy is applied intuitively. For example, a contractor might know that the Aardvark Bidding Company, a notoriously low bidder, wants a particular job very badly. The contractor, then, can choose from several alternatives:

1. He can go ahead and bid the job as usual, even though virtually certain that he will be underbid by Aardvark.
2. He can attempt to underbid Aardvark by lowering his price to a level at which he will make little, if any, profit if he does get the job.
3. He can attempt to maximize his expected profit by bidding at some point which affords him a moderate chance of making a moderate profit.
4. He can ignore the job and look for some other project in which Aardvark is unlikely to be involved.

237

By carefully studying the effect of his bid, both on his chances of getting the job and on the profit that can be achieved were he to receive the job, the contractor will be able to identify the most favorable alternative, which will almost certainly be either the third or fourth of those listed.

Competitive bidding strategies have been applied successfully by many contractors; and, in many cases, an intuitively developed and applied strategy has resulted in highly profitable operations. But an objectively developed strategy, intelligently applied with benefit of a background of management experience, will show results far superior to any method based on intuition. A statistically developed competitive bidding strategy will prove an invaluable supplement to—but *not* a substitute for—informed management judgment.

Most of the examples and figures presented in this chapter are based on the actual experience of operating contractors; the methods of analysis and the techniques of application can be applied by any contractors engaged in competitive bidding.

Information Requirements and Sources

A number of different competitive bidding strategies are employed on every job involving competitive bidding. They may be good bidding strategies, or bad bidding strategies, but nevertheless every bid submitted is the result of some individual's concept of what a bidding strategy should be.

For a good strategy to be applied effectively, however, some information is required regarding the amount and type of competition to be encountered on a job.

Ideally, the contractor should know the names of all competitors on a given job and should have accumulated through experience data regarding their bidding characteristics. In some cases, this ideal situation may exist where perhaps six contractors regularly compete with each other on certain types of work within a limited area. Usually, though, there will be some unknown or unexpected competitive factors involved.

Most public works projects have public bid openings, so the interested and alert contractor has ample opportunity to accumulate data on his competitors for whatever jobs he bids. He should always record all competitors' bids on all projects for which he has prepared a detailed cost estimate.

Projects sponsored by private owners, where bids are not publicly tabulated or reported, may present problems in data collection. A few discreet inquiries regarding competitors and competitors' bids can still yield useful information. Again, any information on competitors' prices should be recorded. At least a contractor who failed to get the job can assume that his bid was not low enough or did not remain low enough to get the job; whoever got the job either bid lower initially or made enough additional concessions after the letting to end up being low bidder. This information, too, is useful.

When the names of specific competitors are not known, simply having some idea of the approximate number of bids that will be submitted on a job will be extremely helpful in determining an appropriate markup. Sometimes the contractor's experience on similar jobs will be sufficient to enable him to estimate the probable number of competitors, based perhaps on the size and type of job and on the current economic condition of his industry.

Another useful source of information is the plan rooms of owners, architects, engineers, and service organizations. A competitor's estimators studying the plans of a forthcoming project provide an obvious and strong indication of that competitor's intention to bid the job. Often the names, or at least the number, of contractors who have checked out plans for a specific project can be obtained. The F. W. Dodge Corporation frequently reports such information. And subcontractors and materials suppliers will know which contractors have asked for prices on a specific job.

And finally, some contractors associations offer a service to their members in maintaining a bid registry where the names of prospective bidders for different jobs are recorded. This service informs member firms of the intentions of their competitors and enables them to select jobs on which competition is relatively less intense. Bid registries do, however, sometimes result in lawsuits and antitrust legislation and must be handled very carefully to avoid such charges.

Developing the Competitive Bidding Strategy

The development of a competitive bidding strategy is a straightforward process, once the general principles are understood and the appropriate data have been collected and analyzed. Again, the goal of a conventional competitive bidding strategy is to find the optimum combination of the profit resulting from getting a job at a given price, and the probability of getting the job at that price.

Several distinct steps are involved in developing the strategy. The first four steps are concerned with preparing the data; the remaining steps are concerned with finding the right bid for the right job. The following seven steps are involved:

1. Tabulate competitors bids on all jobs.
2. Summarize the tabulations for each major competitor.
3. Construct a probability curve for each major competitor.
4. Construct a probability curve for the typical competitor.
5. Identify the competitors involved on the particular job being considered.

6. Determine the probability of being low bidder on the job with any given bid.
7. Compute the expected profit associated with each possible bid, and identify the optimum bid as the markup resulting in maximum expected profits.

Each of these steps will be covered in some detail in the following sections, using examples taken from the operations of several active contractors. Most of the jobs were relatively large, but the job size has little effect on the strategy to be employed.

Bid Tabulation

The Aardvark Bidding Company bids work in the heavy construction field. While operations are conducted throughout the United States, most are concentrated in the Midwest, chiefly in the Missouri and Mississippi River Valleys.

Over a two-year period Aardvark has accumulated information regarding competitors' bids on all jobs for which Aardvark prepared detailed cost estimates. In all, information covering 76 jobs has been obtained. On these jobs, more than 150 different competitors were encountered. Most were encountered only occasionally (nearly 60 percent were encountered on only one job during the two-year period), and only 8 companies were faced on more than 10 jobs during the two years. These 8 included some of the nation's largest and best known general contractors.

Pertinent data regarding the jobs that Aardvark bid during the two-year period are summarized in Table 34.

Table 35 shows the bid tabulations compiled at four different lettings where a group of the same competitors were encountered. On these four jobs, eight different competitors were faced by Aardvark. Of the eight, four were encountered on all four jobs; three competed on three of the four jobs; and one was involved in only one of the jobs.

Summarizing Data on Individual Competitors

Table 36 shows a tabulation of bids pertaining to only one competitor who was faced on 36 jobs during the two-year period. On these 36 jobs, this competitor's bids ranged from 77 percent of Aardvark's estimated direct job cost to 133 percent of the estimated cost. Thus, for Aardvark to have underbid this competitor every time, Aardvark would have had to bid every job at 77 percent of

TABLE 34.   Aardvark Bidding Company: Two-Year Summary of Jobs Bid

Estimated cost of jobs	Number of Jobs	
Under $500,000	26	
$500,000 to $1,000,000	22	
Over $1,000,000	28	
Average size of jobs jobs bid		$1,230,000

Number of competitors per job	Number of Jobs	
1 to 3 competitors	17	
4 to 6 competitors	36	
7 to 9 competitors	13	
10 or more competitors	10	
Average number of competitors per job		5.9 companies

Number of different competitors encountered	Number of Companies	
1 time only	89	
2 to 5 times	46	
6 to 10 times	10	
More than 10 times	8	

cost. Had Aardvark bid each of these 36 jobs at cost, the competitor would have been underbid on 26 jobs, or 72 percent of the total. Obviously, trying to underbid such a competitor on every job would be both ridiculous and diastrous. Nevertheless, money can be made bidding against such a competitor if the bidding is done strategically.

Construction of Probability Curves

The data from Table 36 can be used to construct a probability curve that will indicate, for any given bid, the probability of underbidding this competitor. Figure 66 is the resulting probability curve for this competitor, F. B. Knight Contractors.

The points shown in Figure 66 in each case represent the percentage of bids equal to or higher than a given amount, taken directly from the right-hand column in Table 36. From this curve Aardvark's chances of underbidding the competitor can be easily visualized. For example, a bid placed at 5 percent

TABLE 35. Aardvark Bidding Company: Example of Job Tabulations

Rank	Competitor's Name	Competitor's Bid ($)	Bid as Percentage of Estimated Cost
1	F. B. Knight Contractors	464,920	90.1
2	Gyppo Constructors	472,250	91.7
3	L.Chance Construction Co.	476,745	92.6
4	Able Building Co.	485,510	94.3
5	S. B. Sure Construction Co.	489,055	95.0
6	C. M. Tore & Sons	545,225	106,0
7	L. Gone Contracting Co.	570,630	111.0
8	Aardvark Bidding Co.	591,865	114.9
1	L. Chance Construction Co.	385,735	93.5
2	Able Building Co.	420,890	102.0
3	L. Gone Contracting Co.	430,425	104.5
4	S. B. Sure Construction Co.	432,435	104.6
5	Aardvark Bidding Co.	474,465	115.0
6	C. M. Tore & Sons	487,890	118.2
7	F. B. Knight Contractors	522,205	126.4
1	Able Building Co.	403,340	104.2
2	L. Gone Contracting Co.	405,375	104.9
3	F. B. Knight Contractors	406,385	105.0
4	L. Chance Construction Co.	414,860	107.0
5	S. B. Sure Construction Co.	424,340	109.5
6	C. M. Tore & Sons	425,475	110.0
7	Gyppo Constructors	430,235	111.1
8	Aardvark Bidding Co.	436,650	112.8
1	Gyppo Constructors	431,720	98.5
2	L. Gone Contracting Co.	471,785	107.7
3	F. B. Knight Contractors	474,415	108.1
4	Aardvark Bidding Co.	492,490	112.2
5	Bad Brothers	509,040	116.0
6	Able Building Co.	514,805	117.2
7	L. Chance Construction Co.	573,310	130.7

above cost will be low 54 percent of the time; a 10 percent markup will beat 38 percent of this competitor's bids; and a 15 percent markup will be lower than the competitor's price on 24 percent of the jobs.

Similar data were collected for Aardvark's other major competitors. The resulting probability curves are shown in Figures 67 to 70 for four of these other competitors. In Figure 71 the five probability curves are superimposed, showing how the bidding characteristics of these competitors differ from each other. A markup of 20 percent, for example, pratically eliminates any chance of underbidding C. M. Tore & Sons, but still allows about a 45 percent chance of underbidding F. T. Coop; the same bid results in different probabilities when bidding against the other three competitors.

The Typical Competitor

When the identity of specific competitors on a given job cannot be determined, the concept of a typical competitor is useful. This typical competitor is actually

TABLE 36. **Aardvark Bidding Company: Example of Bid Tabulation for Single Competitor (F. B. Knight Contractors)**

Bid As Percentage of Estimated Cost	Total Number of Bids	Number of Bids Equal or Higher	Percentage of Bids Equal or Higher
77%	1	36	100%
90	1	35	97
92	2	34	94
93	1	32	89
94	1	31	86
95	1	30	83
97	1	29	81
99	2	28	78
100	2	26	72
103	3	24	67
105	2	21	58
106	1	19	53
108	2	18	50
109	1	16	44
110	1	15	42
111	2	14	39
112	1	12	33
113	2	11	31
115	1	9	25
117	2	8	22
121	2	6	17
123	1	4	11
125	1	3	8
127	1	2	6
133	1	1	3

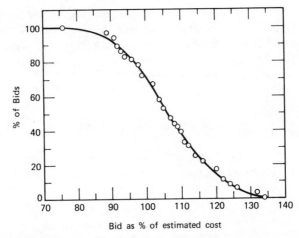

FIGURE 66.   Probability curve for F. B. Knight Contractors.

a composite view of all competitors on all jobs lumped together into one probability curve.

Table 37 summarizes 259 of the bids sumbitted by Aardvark's competitors over a two-year period; Figure 72 plots the probability of Aardvark's being low bidder against a single typical competitor by bidding any given amount above the estimated cost.

Since the probability curve represents the average competitor, it can be used in the absence of any specific knowledge of a given competitor and will usually give satisfactory results when used in this manner.

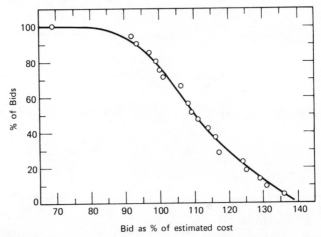

FIGURE 67.   Probability curve for L. Chance Construction Co.

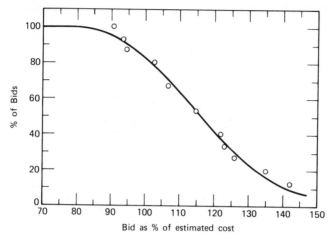

FIGURE 68.   Probability curve for F. T. Coop Builders.

Most jobs will require that the average competitor curve be used in place of actual probability curves for several of the competitors involved, for not all competitors can be identified, and many will be encountered for whom no past data are available. A convenient way of using the average competitor's probability curve is to calculate from it the probability of underbidding different numbers of typical competitors. Table 38 gives the probability of Aardvark's underbidding from 1 to 10 of its typical competitors.

In Table 38 the probability of underbidding one typical competitor is taken directly from the typical competitor's probability curve of Figure 72. To calcu-

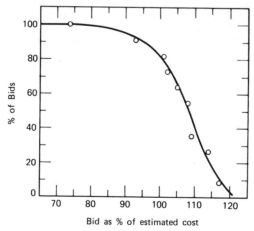

FIGURE 69.   Probability curve for C. M. Tore & Sons.

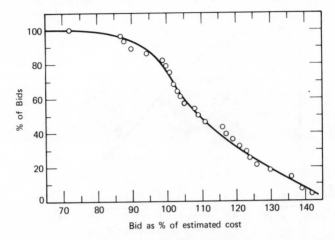

FIGURE 70. Probability curve for S. B. Sure Construction Co.

late the probability of underbidding any number of average competitors, Aardvark need only multiply the probability of underbidding a single competitor by itself that number of times (or, to avoid the calculations, refer again to Table 38). Figure 73 plots Aardvark's probability of underbidding different numbers of competitors with different bids.

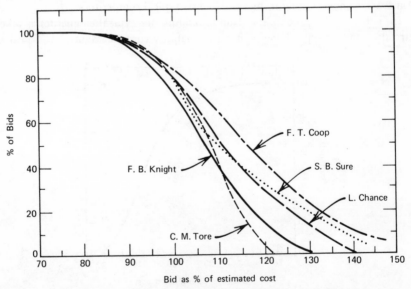

FIGURE 71. Probability curves for five different competitors.

TABLE 37. Aardvark Bidding Company: Bid Tabulation for Typical Competitor[a]

Bid As Percentage of Estimated Cost	Total Number of Bids	Number of Bids Equal or Higher	Percentage of Bids Equal or Higher
50 to 55%	2	259	100.0%
56 to 60	2	257	99.1
61 to 65	2	255	98.4
66 to 70	4	253	97.6
71 to 75	3	249	96.1
76 to 80	7	246	94.9
81 to 85	5	239	92.2
86 to 90	13	234	90.3
91 to 95	18	221	85.3
96 to 100	37	203	78.4
101 to 105	24	166	64.0
106 to 110	30	142	54.8
111 to 115	21	112	43.2
116 to 120	35	91	35.1
121 to 125	15	56	21.6
126 to 130	9	41	15.8
131 to 135	9	32	12.3
136 to 140	3	23	8.9
141 to 145	5	20	7.7
146 to 150	2	15	5.8
151 to 155	4	13	5.0
156 to 160	0	9	3.5
161 to 165	1	9	3.5
166 to 170	2	8	3.1
171 to 175	0	6	2.3
176 to 180	1	6	2.3
181 to 185	0	5	1.9
186 to 190	3	5	1.9
191 to 195	0	2	0.8
196 to 200	0	2	0.8
201 to 205	1	2	0.8
206 to 210	0	1	0.4
211 to 215	0	1	0.4
216 to 220	0	1	0.4
221 to 225	1	1	0.4
226 to 230	0	0	0.0
	259 bids		

[a] Based on a composite of all Aardvark competitors over a two-year period.

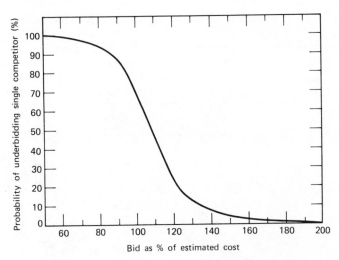

FIGURE 72. Probability curve for typical competitor.

TABLE 38. Aardvark Bidding Company: Probability of Underbidding Different Numbers of Typical Competitors (%)

Markup (%)	Number of Typical Competitors									
	1	2	3	4	5	6	7	9	10	
0	69	47.6	32.9	22.7	15.6	10.8	7.4	5.1	3.5	2.4
1	66	43.6	28.7	19.0	12.5	8.3	5.5	3.6	2.4	1.6
2	64	41.0	26.2	16.8	10.7	6.9	4.4	2.8	1.8	1.2
3	62	38.4	23.8	14.8	9.2	5.7	3.5	2.2	1.4	0.8
4	60	36.0	21.6	13.0	7.8	4.7	2.8	1.7	1.0	0.6
5	57	32.5	18.5	10.6	6.0	3.4	2.0	1.1	0.6	0.4
6	55	30.3	16.6	9.2	5.0	2.8	1.5	0.8	0.5	0.3
7	53	28.1	14.9	7.9	4.2	2.2	1.2	0.6	0.3	0.2
8	50	25.0	12.5	6.3	3.1	1.5	0.8	0.4	0.2	0.1
9	48	23.0	11.1	5.3	2.5	1.2	0.6	0.3	0.1	0.1
10	46	21.2	9.7	4.5	2.1	0.9	0.5	0.2	0.1	0.0
11	44	19.4	8.5	3.7	1.6	0.7	0.3	0.1	0.1	0.0
12	41	16.8	6.9	2.8	1.2	0.5	0.2	0.1	0.0	0.0
13	39	15.2	5.9	2.3	0.9	0.4	0.1	0.1	0.0	0.0
14	37	13.7	5.1	1.9	0.7	0.3	0.1	0.0	0.0	0.0
15	35	12.3	4.3	1.5	0.5	0.2	0.1	0.0	0.0	0.0

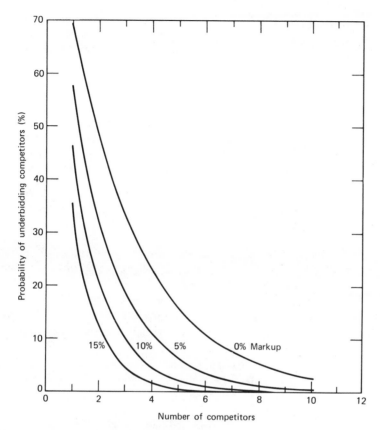

FIGURE 73. Probability of underbidding different numbers of typical competitors.

Identifying the Competitors

Having developed the basic information to be used in determining the optimum bidding strategy for any known situation, the first step in looking at a specific job is to identify the competitors on the job under consideration.

As pointed out in an earlier section, there are a number of ways that competitors can be identified, such as by learning who has checked out plans, whose estimators are inhabiting plan rooms, who has asked for quotations on materials or subcontract work, and who has declared his intention to bid a job.

The Aardvark Bidding Company is interested in four different jobs. Aardvark has learned that the following competitors are planning to bid these four jobs:

Job 1. F. B. Knight, L. Chance, C. M. Tore, and three unknown competitors.
Job 2. F. B. Knight, F. T. Coop, S. B. Sure, and one unidentified competitor.
Job 3. S. B. Sure, F. T. Coop, and C. M. Tore.
Job 4. Five unknown competitors.

Each of the four jobs is estimated at $100,000. If Aardvark's objective were simply to identify which of the four jobs to bid, these estimates could have been made by using shortcut estimating techniques. However, Aardvark in this case intends to bid all four jobs, and the objective is to determine how much to bid on each—or how low to bid to get both job and profit.

The Probability of Being Low Bidder

Having identified the names of known competitors and the number of unknown competitors, the next step is to determine the probability of being low bidder on each job, depending on the bid submitted.

From the probability curve developed for each major competitor, the probability of underbidding that competitor with any given bid can be identified. And from the probability curve developed for different numbers of average competitors, the probability of underbidding any number of average competitors can be found. Combining all the probabilities will give the probability of being low bidder on each job, according to the amount bid.

TABLE 39. Probabilities Associated with Different Bids on Job 1

Markup (%)	Amount Bid ($)	Probability of Underbidding (%)				
		F.B. Knight	L. Chance	C.M. Tore	3 Unknown Competitors	All Six
0	100,000	70	78	83	33	15.0
1	101,000	67	76	82	29	12.1
2	102,000	64	74	80	26	9.8
3	103,000	61	71	78	24	8.1
4	104,000	58	68	76	22	6.6
5	105,000	54	65	74	19	4.9
6	106,000	50	62	72	17	3.8
7	107,000	47	59	70	15	2.9
8	108,000	44	56	68	13	2.2
9	109,000	41	53	66	11	1.6
10	110,000	38	51	64	10	1.2

TABLE 40. Probabilities Associated with Different Bids on Job 2

Markup (%)	Amount Bid ($)	Probability of Underbidding (%)				
		F.B. Knight	F.T. Corp.	S.B. Sure	1 Unknown Competitor	All Four
0	100,000	70	82	77	69	30.5
1	101,000	67	80	74	66	26.2
2	102,000	64	77	70	64	22.1
3	103,000	61	74	66	62	18.5
4	104,000	58	70	63	60	15.4
5	105,000	54	66	60	57	12.2
6	106,000	50	61	57	55	9.6
7	107,000	47	56	54	53	7.5
8	108,000	44	52	52	50	5.9
9	109,000	41	46	50	48	4.5
10	110.000	38	40	48	46	3.4

Table 39 shows the probability of being low bidder on job 1 with bids ranging from $100,000 (the estimated direct cost) to $110,000 (a 10 percent markup). Tables 40, 41, and 42 show comparable data pertaining to jobs 2, 3, and 4. The probabilities of underbidding the individual competitors are taken from Figures 32 to 35, and the probability of underbidding different numbers of typical competitors comes from Table 38.

TABLE 41. Probabilities Associated with Different Bids on Job 3

Markup (%)	Amount Bid ($)	Probability of Underbidding (%)			
		S.B. Sure	F.T. Corp.	C.M. Tore	All Three
0	100,000	77	82	83	52.4
1	101,000	74	80	82	48.6
2	102,000	70	77	80	43.1
3	103,000	66	74	78	38.1
4	104,000	63	70	76	33.5
5	105,000	60	66	74	29.3
6	106,000	57	61	72	25.0
7	107,000	54	56	70	21.2
8	108,000	52	52	68	18.4
9	109,000	50	46	66	15.2
10	110,000	48	40	64	12.3

TABLE 42. Probabilities Associated with Different Bids on Job 4

Markup (%)	Amount Bid ($)	Probability of Underbidding Five Unknown Competitors
0	100,000	15.6
1	101,000	12.5
2	102,000	10.7
3	103,000	9.2
4	104,000	7.8
5	105,000	6.0
6	106,000	5.0
7	107,000	4.2
8	108,000	3.1
9	109,000	2.5
10	110,000	2.1

In Table 39 Aardvark is found to have a 15.0 percent chance of being low bidder on job 1 if the bid is submitted at cost, with no markup ($.70 \times .78 \times .83 \times .33$). The chances of getting the job decrease as the bid increases, down to only a 1.2 percent chance of being low bidder with a $110,000 bid ($.38 \times .51 \times .64 \times .10$).

On job 2 (Table 40) a bid placed at cost would be expected to win 30.5 percent of the jobs. A 5 percent markup would be successful 12.2 percent of the time, and a 10 percent markup would be low only 3.4 percent of the time against these competitors.

Job 3, with only three competitors involved, offers a greater likelihood of success. As shown in Table 41, there is a 52.4 percent chance that a bid placed at cost would be low; a 29.3 percent chance for a 5 percent markup to be successful; and a 12.3 percent chance that a 10 percent markup would win the job.

On job 4, against five unknown competitors, Aardvark can expect to be low bidder only 15.6 percent of the time, even when bidding at cost. The probability of success is reduced to 6 percent with a 5 percent markup, and 2.1 percent with a 10 percent markup.

Expected Profits on Each Job

Table 39 through 42 show the probability of being low bidder on each of the four jobs against different competition with any given bid.

But profit—not probability—is the key to successful bidding. Simply getting jobs is not enough; they must be won at a profit. The next step, then, is to determine the expected profit associated with each bid on each job.

On job 1, Aardvark can expect to be low bidder 12.1 percent of the time by bidding 1 percent above cost. This would result in making 1 percent (or $1000), 12.1 percent of the time, an average of $121 for each time such a bid is submitted in this type of situation. Or, Aardvark can bid with a 10 percent markup and be successful 1.2 percent of the time, an average of $120.

By thus combining the probability of being low bidder with any given bid, with the amount of profit to be made should that bid be successful, Aardvark can easily determine exactly at what point the bid should be placed to result in the maximum possible long-run profits. Table 43 summarizes the expected profit for the different bids on each of the four jobs.

On job 1, expected profit rises from zero at a bid placed at cost (a 15 percent chance of making nothing) to a high of $264 at a 4 percent markup (corresponding to a 6.6 percent chance of making 4 percent of $4000), then back down to $120 at a 10 percent markup (where there is a 1.2 percent chance of making $10,000. The optimum bid in this case, then, is $104,000. A higher bid will pick up more profit on the jobs won, but will not make up for the jobs lost by the additional markup. On the other hand, a low bid will win more jobs, but even so, the resulting profit will be less at the lower markup. Figure 74 shows how the expected profit on job 1 varies with the percentage markup.

TABLE 43.   Summary of Expected Profit on Jobs 1 to 4

Markup (%)	Expected Profit ($)			
	Job 1	Job 2	Job 3	Job 4
0	0	0	0	0
1	121	262	486	125
2	196	442	862	214
3	243	555	1,143	276
4	264[a]	616[a]	1,340	312[a]
5	245	610	1,465	300
6	228	576	1,500[a]	300
7	203	525	1,484	294
8	176	472	1,472	248
9	144	405	1,368	225
10	120	340	1,230	210

[a] Optimum bid.

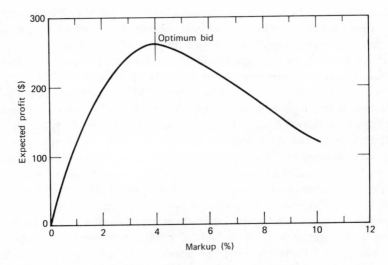

FIGURE 74.   Expected profit curve for job 1.

Job 2 offers similar alternatives. Again, a bid at cost will result in no profit, regardless of the number of jobs won. The maximum profits of job 2 occur at a markup of 4 percent, where there is a 6.6 percent chance of being low bidder, for an expected profit of $616 (15.4 percent of $4000). Even though the optimum bid on job 2 has the same markup as the optimum bid for job 1—4 percent—job 2 offers more than twice the amount of expected profit as job 1 because of the better chance of getting the job at this markup. Figure 75 shows the shape of the expected profit curve for job 2.

Job 3 offers both the highest optimum bid and the highest expected profit of the four jobs, since the probability of getting the job with any given markup is higher than for any of the other three jobs. Here, the optimum bid occurs at a markup of 6 percent, where a 25 percent chance of being low bidder results in an expected profit of $1500. Even a bid of $110,000 on this job will result in expected profits of $1230 (a 12.3 percent chance of making $10,000), nearly twice as much as the maximum expected profit offered by job 2 and more than four times the maximum of job 1. The expected profit curve for job 3 is shown in Figure 76.

Job 4 has better profit potential than job 1, but not as good as jobs 2 or 3. The optimum bid on job 4 is at $104—again a 4 percent markup, as in jobs 1 and 2. But the expected profit amounts to $312, resulting from a 7.8 percent chance of making $4000. The expected profit then tapers off to $210 at a 10 percent markup. The expected profits associated with each bid are shown in Figure 77.

If Aardvark were unable to bid all four jobs, maximum total profits could be achieved by bidding job 3 first, followed by jobs 2 and 4. Job 1 offers the least attractive opportunity in terms of profit potential, as might be expected since it involves the largest number of competitors. In fact, the relative desirability of all four of these jobs varies inversely with the number of competitors encountered.

Actually, three of the jobs—1, 2, and 4—offer so little opportunity for profit that the cost of preparing estimates and bidding the jobs probably cannot be justified. On job 1, for example, if the cost of estimating the job exceeded $264—the maximum expected profit for the job—Aardvark would end up losing money by even bidding this type of job. Better results could be obtained in the long run by ignoring such jobs and by searching out more profitable opportunities.

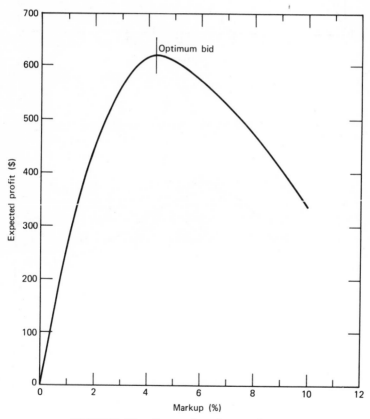

FIGURE 75. Expected profit curve for job 2.

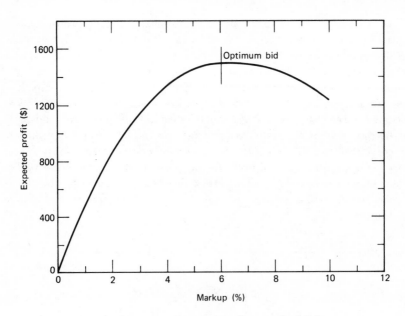

FIGURE 76. Expected profit curve for job 3.

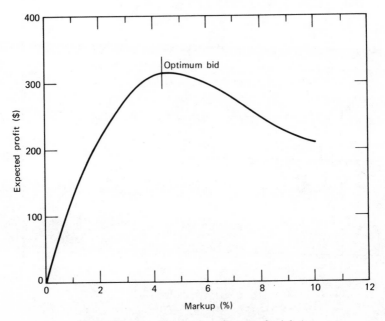

FIGURE 77. Expected profit curve for job 4.

The Optimum Bid Against Unknown Competitors

When bidding against unknown competitors, against competitors for whom insufficient data are available, or even against a large number of known competitors, data for the typical competitor can usually be submitted for more specific data regarding individual competitors, still with good results. When this method is used, the optimum markup is assumed to be a direct function of the number of bidders involved.

The probability curve for Aardvark's typical competitor is shown in Figure 72. This curve can be used as the basis for calculating expected profits when bidding agaist different numbers of typical competitors.

Figure 78 shows how Aardvark's optimum bid and expected profit vary when bidding against one, two, and four typical competitors.

As is evident from Figure 78, an increase in the number of competitors decreases both the optimum bid and the expected profit on the job. When bidding against only one typical competitor, the optimum bid is at a markup of about 15 percent, where the probability of success is 35 percent, and the expected profit reaches its maximum of 5.25 percent. Against two typical competitors, the optimum bid drops to 11 percent with a maximum profit expectation of about 2.1 percent. Four competitors on a job reduce the optimum bid to

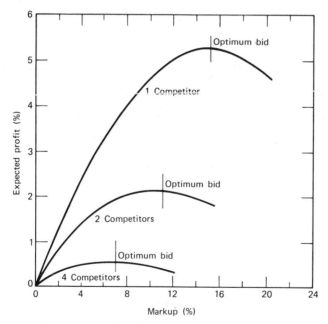

FIGURE 78. Expected profit against 1, 2, and 4 typical competitors.

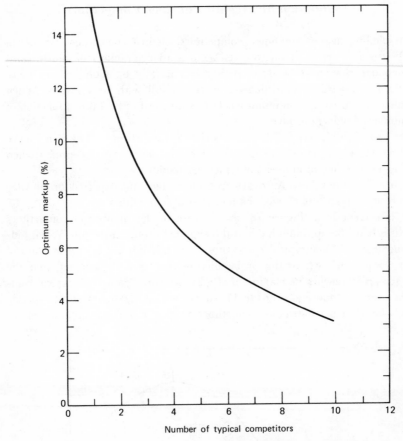

FIGURE 79.   Effect of number of typical competitors on optimum bid.

7 percent, where a 7.9 percent chance of a winning bid results in an expected profit of 0.6 percent. The effect of the number of typical competitors on the optimum bid is shown in Figure 79.

This type of analysis—using the number of competitors on a job as the sole basis for defining the optimum markup—admittedly lacks a great deal of refinement. But the results of using these markups can sometimes be spectacular, and they will invariably be better than what can be achieved by any "hunch" method.

Summary

Many information sources are available to the contractor in obtaining the necessary data to develop a working strategy for his operation. After the necessary data have been accumulated, development of the competitive bidding strategy is a straightforward process, involving several distinct steps. The first step is to tabulate competitors' bids on all jobs, next, to tabulate the bids individually for each major competitor, then, to construct probability curves both for individual competitors and for the typical, or average, competitor, showing the chances of underbidding a particular competitor by bidding any specified amount. After the competitors to be encountered on a particular job have been identified, the probability of being low bidder on the job with any given bid can easily be calculated, and the bid resulting in the highest possible expected profit can be found. If specific competitors cannot be determined prior to the time that bids are submitted, the characteristics of the typical competitor can be substituted with good results. In any case, the optimum bid and the expected profit will vary inversely with the number of bidders on the job; in other words, the profits to be gained from any job will decrease as the intensity of competition increases, and the optimum bid will reflect this fact.

19

The Databid System

Over the past two decades, there has been a great deal of controversy and disagreement regarding the applicability and effectiveness of various theoretical approaches to competitive bidding. Several models are available to choose from, but no amount of discussion or research will change one basic truth: a good model must yield good results when put into practice.

The Databid System presented in this chapter is offered as a simple, practical, and proven approach to developing an effective data-based competitive bidding strategy.

It is an approach, a method, but *not* a theory. It is entirely factual, and whether or not the contractor chooses to put its findings into practice, it still provides much valuable information. The contractor can easily determine exactly how much profit he is sacrificing by not following it.

Objective of the Databid System

The overall objective of the Databid System is simple: to make as much profit as possible on each job bid, based on the job characteristics and the prevailing competitive situation.

Based on the results achieved by a number of different contractors operating in different fields and in different geographic areas, a contractor can, by using the Databid System, make about half as much gross profit as he could if he knew all his competitors bids in advance.

260

The Databid Approach

In the Databid System, the competitive information described in preceding chapters is grouped into categories based on job size and the number of competitors encountered on each job. The specific breakdown will vary among different contractors; as few as three or as many as seven classifications may be used for each of the two variables, giving a total of from 9 data groups (three job size groups for each of three competitor groups) to 49 data groups (for seven job size and seven competitor groups). Figure 80 illustrates how these groupings might be set up.

The actual number and size of data groupings will depend on the characteristics of the contractor's jobs and on the amount of data available. Ideally, there will be some data to enter in each category, or "cell."

	Number of Competitors		
Job Size	1 or 2	3 to 5	6 or More
Under $10,000			
$10,000 to $30,000			
Over $30,000			

	Number of Competitors						
Job Size	1	2	3	4	5	6	7 or More
Under $20,000							
$20,000 to $50,000							
$50,000 to $100,000							
$100,000 to $200,000							
$200,000 to $500,000							
$500,000 to $1,000,000							
Over $1,000,000							

FIGURE 80.   Grouping job data for Databid analysis.

About five jobs per cell is a good number to start with, although more is better. Also, it is preferable to work with recent jobs, those bid within the past year or two.

This means that if 50 jobs were bid during the last year, the matrix should contain about 10 cells, while 100 jobs would provide enough data for about 20 cells (perhaps a 4 × 5 or 5 × 4 matrix).

Essentially, what will be done with the data after it has been grouped will be to develop a separate bidding strategy for each cell, based on the data contained only in that cell.

For example, suppose that five jobs were bid in the $20,000 to $50,000 range against three competitors. By extracting these five jobs from the information system described in Chapter 15, the following information could easily be obtained:

Job Number	Lowest Competitor's Markup (%)
1	9.2
2	14.6
3	3.4
4	7.7
5	18.3

Identifying the optimum bid on these five jobs, then, entails only a simple analysis to determine the one markup that would have had the best results had it been applied to all five jobs falling in this particular classification. Table 44 shows the results of this analysis.

TABLE 44.   Identifying the Optimum bid

Markup (%)	Number of Jobs Won	Expected Profit (%)
0	5	0
3	5	15
4	4	16
7	4	28
8	3	24
9	3	27
10	2	20
14	2	28
15	1	15
18	1	18
19	0	0

The "expected profit" column in Table 44 is calculated by multiplying the percentage markup by the number of jobs won at that given markup. In the example, it is apparent that this contractor would not have wanted to bid with a markup below 6 percent or above 14 percent. His best choices would have been to take either four jobs at a 7 percent markup, or two jobs at a 14 percent markup; 3 jobs at 9 percent would be almost as good.

Looking back, had he known every competitor's bid in advance, a contractor could have taken all five jobs at markups of 9.1, 14.5, 3.3, 7.6, and 18.2 percent. Not knowing these competitive bids in advance, the two jobs taken at a 14 percent markup would have given him a little more than half as much total profit as the most he could have made even if he had known.

Applying the 14 percent optimum markup on future jobs bid in the same size category against the same number of competitors will not, of course, guarantee similar results. Nevertheless, the results of following the guidelines established by this analysis are likely, in practice, to be better than anything else he can do without prior knowledge of his competitors' intentions.

As the same approach is applied to each cell in the matrix, the contractor will begin to see definite patterns emerge.

In general, the matrix will be expected to show the highest optimum markups on small jobs bid against few competitors, and the lowest optimum markups on large jobs bid against many competitors. In between these two extremes, the results are far less predictable.

The unpredictability of results, though, poses no real problem because the whole objective is to determine *what* has happened, and nothing in particular should be expected prior to the analysis.

While this approach may be irritating to a theoretician, it works nevertheless.

Application of the Databid System

Since the Databid System is not a theory, there can be no theoretical discussion of it. Its sole purpose is to find the best or optimum markup for each type of job bid by a particular contractor based on his own experience.

The analysis will turn out different for every contractor, so no general conclusions can be drawn that will apply universally, except one: better profits can be made with it than without it.

The following example is based on the experience of a special trades contractor. This contractor bid 74 jobs that were let at above his estimated direct cost during the preceding year. The jobs ranged in size from $1200 to $743,400, and were bid against as few as 1 and as many as 10 competitors. Table 45 summarizes the job sizes, number of competitors on each, and the lowest competitor's markup.

TABLE 45. Summary of Bid Tabulations

Job Number	Job Size ($1000)	Number of Competitors	Lowest Competitor's Markup (%)	Job Number	Job Size ($1000)	Number of Competitors	Lowest Competitor's Markup (%)
1	26.3	2	29	38	36.3	5	9
2	2.8	1	10	39	7.0	3	7
3	6.1	1	31	40	34.4	5	4
4	3.0	5	20	41	2.2	1	5
5	1.2	1	40	42	9.4	3	33
6	23.5	1	26	43	7.5	1	41
7	2.0	1	7	44	2.9	1	27
8	11.4	1	14	45	44.8	2	7
9	13.6	1	21	46	192.9	5	15
10	125.4	2	15	47	9.8	2	12
11	1.4	2	7	48	24.0	4	8
12	80.9	5	28	49	18.7	1	28
13	743.4	3	2	50	3.7	1	37
14	64.3	10	10	51	12.8	5	16
15	3.3	1	24	52	3.5	3	9
16	69.3	2	12	53	20.4	3	3
17	5.9	1	9	54	7.4	2	4
18	229.9	6	5	55	3.7	3	34
19	4.9	3	16	56	97.7	7	1
20	3.2	6	9	57	12.5	2	7
21	22.3	2	3	58	6.1	2	16
22	25.1	3	6	59	464.2	4	2
23	2.4	2	21	60	86.5	7	12
24	43.2	6	1	61	2.2	1	39
25	2.1	1	29	62	4.2	1	44
26	1.8	2	50	63	3.5	1	22
27	29.9	2	3	64	19.9	2	42
28	16.2	1	37	65	4.4	2	28
29	8.2	2	15	66	357.5	5	2
30	2.1	2	42	67	3.0	1	53
31	16.5	4	7	68	5.1	2	57
32	2.4	1	32	69	2.9	5	9
33	8.6	3	32	70	23.6	3	3
34	96.2	1	33	71	20.9	1	35
35	2.1	2	8	72	5.4	1	39
36	8.3	2	24	73	13.8	1	12
37	16.1	1	9	74	3.3	1	51

The information shown in Table 45 was transferred onto edge-notched data cards as described in Chapter 15, to make the subsequent analysis easier and to provide for continual updating of the data. Figure 81 shows the distribution of the number of competitors encountered on each.

With 74 jobs to analyze, it was decided to use a 4 × 4 matrix, grouping the data into four job size categories and four competitor categories. Figure 82 shows the categories chosen in this case and the number of jobs falling into each category.

The data could just as easily have been grouped in several other ways, but this seemed as convenient and useful a way as any.

With the job categories thus established, the next step is to find the optimum bid in each cell of the matrix for which there are sufficient data. Table 46 shows how this was done for the under-$5000 projects bid against a single competitor, of which there were 14.

From among the 14 jobs in this cell, the lowest competitor's markups shown in Table 45 ranged from 5 percent (on job 41) to 53 percent (on job 67.). By bidding with a 4 percent markup on all jobs, the contractor would have been low bidder on all 14, as shown in Table 46. A markup of 9 percent would have

Job Size ($1000)	Number of Competitors										
	1	2	3	4	5	6	7	8	9	10	Total
0 to 2	2	2	–	–	–	–	–	–	–	–	4
2 to 5	12	4	3	–	2	1	–	–	–	–	22
5 to 10	4	6	3	–	–	–	–	–	–	–	13
10 to 20	6	2	–	1	1	–	–	–	–	–	10
20 to 50	2	4	3	1	2	1	–	–	–	–	13
50 to 100	1	1	–	–	1	–	2	–	–	1	6
100 to 200	–	1	–	–	1	–	–	–	–	–	2
200 to 500	–	–	–	1	1	1	–	–	–	–	3
500 to 1000	–	–	1	–	–	–	–	–	–	–	1
Total	27	20	10	3	8	3	2	0	0	1	74

FIGURE 81. Tabulation of jobs by category showing the frequency of encountering different numbers of competitors on jobs of various sizes.

Job Size ($1000)	Number of Competitors				
	2	2	3 to 5	6 or More	Total Number of Jobs
0 to 5	14	6	5	1	26
5 to 20	10	8	5	0	23
20 to 100	3	5	7	4	19
Over 100	0	1	4	1	6
Total Number of Jobs	27	20	21	6	74

FIGURE   82. Breakdown of job characteristics used for development of bidding strategy.

eliminated only two jobs and taken the other 12, while a 21 percent markup would have been good on 11 of the 14 jobs.

Going through the list of jobs this way, it will be found that the best bid of all would have been a 36 percent markup, taking seven jobs. More jobs could be won with a lower markup and fewer jobs with a higher markup, but the 36 percent markup will yield the highest possible profits. The 36 percent markup, therefore, represents the optimum bid for this class of job.

The same type of analysis is shown in Table 47 for jobs falling in the $20,000 to $50,000 range, bid against six or more competitors. Here, there were only four jobs, with the lowest competitor's markups falling between 1 and 12 percent. Table 47 shows the analysis based on these four jobs.

In Table 47 the expected profit reaches its maximum at a nine percent markup, where two of the jobs can be won. This, then, is the optimum markup for jobs of this type.

By using the same approach in each of the other cells that contain enough jobs to work with, the optimum bids can be determined for each combination of job size and number of competitors. Figure 83 shows the results of these separate analyses.

The table shown in Figure 83 is the contractor's most important single guide to increasing his profits. It tells him how much markup to apply to any job, and, along with the other information already developed, can indicate to him which jobs to bid.

Assigning Priorities

The analysis can be carried a step further by determining the relative attractiveness of projects falling in each category. All the information is already there; it need only be interpreted.

For example, the 36 percent markup on small jobs (under $5000) bid against a single competitor will win half the jobs bid in that category. This represents a 50 percent chance of making a 36 percent gross profit, or an expected profit of 18 percent per job bid.

TABLE 46. Identification of Optimum Bid for Jobs $5000 and Under with One Competitor

Markup (%)	Number of Jobs Won	Expected Profit (%)
0	14	0
4	14	56
5	13	65
6	13	78
7	12	84
9	12	108
10	11	110
21	11	231
22	10	210
23	10	230
24	9	216
26	9	234
27	8	216
31	8	248
32	7	224
36	7	252
37	6	222
38	6	228
39	4	156
40	3	120
43	3	129
44	2	88
50	2	100
51	1	51
52	1	52
53	0	0

**TABLE 47.   Identification of Optimum Bid for
$20,000 to $50,000 Jobs with Six or More
Competitors**

Markup (%)	Number of Jobs Won	Expected Profit (%)
0	4	0
1	2	2
2	2	4
3	2	6
4	2	8
5	2	10
6	2	12
7	2	14
8	2	16
9	2	18
10	1	10
11	1	11
12	0	0

Similarly, the 18 percent markup on $20,000 to $50,000 jobs bid against six or more competitors will be successful on half the jobs: thus, there is a 50 percent chance of making 18 percent for an expected profit of 9 percent on those jobs.

Similar calculations can be made for each category of job characteristics, resulting in the profit expectations shown in Figure 84.

Then, ranking the various categories of jobs by combining the gross amount of profit to be made, the chances of getting the job with a specified markup, and

Job Size ($1000)	Number of Competitors			
	1	2	3 to 5	6 or More
0 to 5	36	41	45	–
5 to 20	27	23	31	–
20 to 100	25	28	27	9
Over 100	–	–	14	–

FIGURE 83.   Optimum markups by job size and number of competitors.

Job Size ($1000)	Number of Competitors			
	1	2	3 to 5	6 or More
0 to 5	18.0	13.7	27.0	–
5 to 20	13.5	8.7	12.4	–
20 to 100	25.0	5.6	3.9	4.5
Over 100	–	–	3.5	–

FIGURE 84. Expected percentage profit per job bid.

the profit expectation applying to that category of job, each cell can be assigned a priority ranking. These rankings are shown in Figure 85.

The projects falling in Priority Group I ($20,000 to $100,000 in size, bid against only one competitor) will, for example, yield higher profits than those in Priority Group II (jobs over $100,000, bid against 3 to 5 competitors). In this case, the higher optimum markup applied to the Group I jobs (25 percent) will more than make up for their smaller size. This chart, then, tells the contractor that the most potentially profitable type of jobs for him to bid are those falling in the $20,000 to $100,000 range that attract only one other bidder. Given a choice, these are the projects he should seek.

On the other hand, small projects will not contribute much to his overall operation despite the relatively high markups accompanying these jobs, *unless* he chooses to "specialize" in these jobs and aggressively seek out many more of them than he has in the past. The priorities shown in this figure are based on the same mix of jobs bid in the past.

Job Size ($1000)	Number of Competitors			
	1	2	3 to 5	6 or More
0 to 5	X	XI	IX	–
5 to 20	VI	VIII	VII	–
20 to 100	I	III	V	IV
Over 100	–	–	II	–

FIGURE 85. Assigning priorities to job types.

A Word of Warning

It is a common practice in research to discard data that do not fit the researcher's preconceived notions. In the analysis just completed, for example, dropping just a few of the jobs out of Table 45 would have made the results conform more closely to what would theoretically be expected.

Experience has shown, however, that actual results seldom conform to theoretical expectations. Nothing is gained—and something is certainly lost—by attempting to make research look like the data are proving a theory.

The data in the preceding section do not prove anything; they simply represent a specific contractor's experience. Every contractor who undertakes a study of this type should enter into it with a completely open mind and be willing to accept the results however they come.

This is why the Databid System is not presented as a theory, and only a few general observations are offered regarding its results. The contractors who have used this approach have achieved profits amounting to 40 to 60 percent of the most they could have made by knowing all of their competitor's bids in advance.

The Databid System is offered only as another tool to aid the contractor in making a good decision at the right time.

Summary

The Databid System offers a practical approach to developing a competitive bidding strategy by organizing and analyzing data on past jobs. In the Databid System, jobs are grouped into categories based on job size and the number of competitors encountered on each job. An optimum bid is then calculated for each combination of job size and number of competitors for which there are sufficient data. These optimum bids will reveal competitive patterns, enabling the contractor to fill in the gaps where data are not available on a judgment basis. Also, appropriate priorities can be assigned to the various job classifications based on the job sizes, optimum markups, and expected profits. Experience has shown that contractors employing the Databid System can achieve gross profits amounting to 40 to 60 percent of the most they could have made by knowing all competitors' bids in advance.

20

Results: The Measure of Success

We have only to explain failures;
success speaks for itself.

HARRY HUNTT RANSOM

The two essential elements in the overall company success plan are strategy and tactics. Strategy involves knowing what to do. Tactics involves doing it.

There is but one real measure of the effectiveness of any competitive strategy: profits. If the profits resulting from using a particular strategy are higher than can be achieved by any other available means, then it is a good strategy; if not, the strategy is worthless.

There is no place in the contracting field for wishy-washy management. The objective of being in business is to make a profit. And whatever strategy helps to achieve the maximum profits possible under existing conditions should be employed.

Failure to use the new management tools is at least as harmful—and frequently more harmful—than failure to use modern equipment or to employ the most efficient work methods. Inefficient operations on the part of management are every bit as inexcusable as inefficient operations in the field.

And yet many contractors who would immediately discharge an employee for performing his work in a sloppy or inefficient manner are guilty of handling their own business operations on a haphazard, intuitive basis with most decisions based on hunches.

For the strategic plan to be successful, it need only be implemented, controlled, and continually revised in the face of changing competitive conditions.

Selecting the Preferred Strategy

The procedures for finding the optimum bid that are discussed in the preceding chapters are aimed at maximizing profits—finding the bid that will in the long run result in the highest possible profits. The contractor may, however, have objectives other than maximizing profits, including the following:

1. Minimizing the possibility of loss.
2. Maintaining some prescribed level of return on investment.
3. Achieving a specified share of the total market.
4. Increasing the likelihood of obtaining a particular job.

Regardless of his specific objectives, the competitive data that have been developed will prove helpful to the contractor in attaining his desired objectives. Simply recognizing the probability of success associated with any bid on any job, and the amount of profit associated with each bid, the contractor can not only identify his optimum bid in terms of gross profits, but can also evaluate the expected outcome of any other bid and its compatibility with any chosen objectives.

For example, the optimum bid on a job may be found by calculations to occur at a markup of 3 percent above estimated direct job costs. If the contractor feels that a minimum markup of 10 percent is required in his operation, he has several alternatives:

1. He can go ahead and bid the job at a 3 percent markup and reevaluate his company's objectives in terms of the existing competitive conditions.
2. He can bid the job with a 10 percent markup, recognizing that his expected profit will be substantially less than if he were to bid it at 3 percent.
3. He can ignore the job, thus saving his estimating and bidding costs, and look for another job offering a profit potential more in line with his present objectives.

The selection of a preferred course of action, then, depends on exactly what the contractor hopes to achieve. But the effect of a competitive decision can, and should, be appraised objectively.

Whatever course of action the contractor pursues, the knowledge gained through the development of competitive information will help him anticipate the probable results of his actions and decisions with a degree of precision that would not otherwise be possible.

Evaluating the Strategy's Effectiveness

Since no changes in estimating or operating procedures are required in applying a competitive bidding strategy, the strategy can be implemented as soon as sufficient data are available, after management is completely convinced of the strategy's value and usefulness and is aware of its limitations.

However, the strategy can—and should—be tested without risk prior to its actual use in competition. This can be done simply by bidding jobs in the usual way, but at the same time determining the optimum bid for the job as indicated by statistical means. By thus comparing the profits actually realized with those that would have resulted had the strategy been employed, an objective and realistic measure of the proposed strategy's effectiveness can be obtained.

After this comparison between actual and potential profits has been made for a sufficient period of time, some indication will have been gained of the proposed strategy's probable effect on the company's overall operations, including its impact both on profits and on the volume of work obtained.

The sales volume resulting from using a bidding strategy can change in either direction. If the previous level of markups has been too low, using a higher markup will most likely result in a lower sales volume accompained by higher profits. Had markups been too high, the volume will probably increase when markups are lowered to an optimum level, again with an increase in profits.

In either case, then, whether more jobs are taken at a lower markup or fewer jobs at a higher markup, profits will increase when a properly designed competitive bidding strategy is implemented.

One Contractor's Experience

As pointed out previously, the effect of a competitive bidding strategy on profits and sales volumes will depend on the contractor's current methods of operation. The following example reflects the actual experience of a large general contractor in the heavy construction field.

The practice of this contractor was, essentially, to bid all jobs at a fixed markup. The management's philosophy was that a certain mimimum margin was required for their business, that a smaller margin was unacceptable, and that a larger margin was neither justified nor possible. Consequently, this contractor's operations had deteriorated almost to the point of complete financial disaster.

The procedure followed in developing a competitive bidding strategy for this contractor was to use all available information on jobs bid during the past two years. Fortunately, due to the nature of the work—mostly public works projects—the names and bids of nearly all competitors were available on all the jobs bid. Also, the company had maintained careful records of its own estimated costs on jobs bid and actual costs on jobs received; these records showed the company's estimates to be sufficiently accurate and reliable, so that estimated costs could be assumed equal to the actual costs.

All bids were considered in the analysis. They ranged from less than half to more than double the company's cost estimates. From these data the characteristics of "average" or "typical" bidders were developed. When specific competitors were encountered frequently—on 10 or more jobs during the two-year period—their characteristics were determined individually.

From the beginning it was apparent that on jobs involving a large number of bidders the individual characteristics of specific competitors were obscured by the numerous other bidders. Therefore, only the information developed for the typical competitor was used in determining optimum bids; the optimum bid on any given job was thus assumed to be a direct function of the number of bidders. On this basis, the following amounts were added to estimated costs to arrive at the optimum bids on subsequent jobs:

Number of Competitors	Markup (%)	Number of Competitors	Markup (%)
1	16.0	7	4.1
2	10.0	8	3.7
3	7.8	9	3.3
4	6.4	10	3.1
5	5.5	11 and over	3.0
6	4.1		

No changes were made in estimating or operating procedures; the only change in the established methods of doing business was in the use of a variable markup.

This contractor bid 36 jobs the following year. Table 48 describes these jobs in terms of estimated costs and the number of competitors—the only information required of a job in using the competitive bidding strategy.

A total of 238 competitors was encountered on these 36 jobs, an average of nearly 7 per job. The total estimated cost of all jobs bid was $48,485,000, slighly more than $1,300,000 per job.

Using the variable markups, this contractor won 16 of the 36 jobs he bid. Of the remaining 20 jobs, 13 were let at a price less than his estimated direct costs for the work. Table 49 summarizes the results of the year's operations.

Jobs totaling $21,715,000 were received during the year. The sum of the second-low bids on these same 16 jobs was $22,387,000, meaning that $672,000—an average of $42,000 per job—was left "on the table." This amount represents an average spread of approximately three percent, certainly not an unreasonable amount for the jobs involved.

Hindsight might show that some other combinations of variable markups would have proved even more profitable. But a primary requisite of a competi-

TABLE 48. Jobs Bid Using Competitive Bidding Strategy

Job Number	Number of Competitors	Estimated Direct Cost ($1000)
1	5	961
2	2	1,361
3	11	808
4	2	729
5	5	160
6	7	556
7	7	387
8	6	299
9	3	2,952
10	5	462
11	5	2,151
12	10	2,261
13	5	93
14	4	992
15	6	150
16	3	1,711
17	11	3,610
18	6	525
19	2	1,086
20	7	1,504
21	3	2,770
22	7	499
23	5	450
24	5	414
25	9	1,901
26	6	1,465
27	6	2,049
28	9	1,314
29	9	4,070
30	14	1,009
31	9	1,684
32	13	3,728
33	11	504
34	5	3,472
35	12	334
36	3	64
Total	238	48,485
Average	6.6	1,347

TABLE 49. **Results of Using Competive Bidding Strategy**

Job Number	Optimum Markup (%)	Optimum Bid ($1000)	Lowest Competitor's Bid ($1000)
1	5.5	1014[a]	1082
2	10.0	1497	1226
3	3.0	832	713
4	10.0	802[a]	824
5	5.5	169	139
6	4.1	579	523
7	4.1	403[a]	403
8	4.7	313	304
9	7.8	3182[a]	3353
10	5.5	487[a]	504
11	5.5	2269	2232
12	3.1	2331	2168
13	5.5	98[a]	106
14	6.4	1055	931
15	4.7	157	154
16	7.8	1844	1749
17	3.0	3718[a]	3732
18	4.7	550	482
19	10.0	1195	1133
20	4.1	1566[a]	1680
21	7.8	2986	2651
22	4.1	519[a]	545
23	5.5	475	451
24	5.5	437[a]	456
25	3.3	1964[a]	1976
26	4.7	1534[a]	1555
27	4.7	2145	1929
28	3.3	1357	1279
29	3.3	4204	4169
30	3.0	1039	928
31	3.3	1740[a]	1860
32	3.0	3840	3424
33	3.0	519[a]	569
34	5.5	3663[a]	3664
35	3.0	344	329
36	7.8	69[a]	78

[a] Low bidder.

tive bidding strategy is that it be developed *before*—not after—the job is let. Hindsight is infallible.

Variable Versus Fixed Markups

The preceding example showed only the results of using the theoretically derived optimum markups. The results were obviously extremely successful. Even so, an analysis of what might have happened had other markups been used should provide some interesting comparisons.

Unless the contractor bids only one type of job against a fixed number of competitors in a completely stable market—a situation facing few contractors— he will do much better in the long run to vary his markups according to his competition. If he generally faces a relatively small number of competitors on any given job, perhaps six or less, and if he frequently encounters the same competitors, he can probably raise his profits substantially by considering both the number and the individual characteristics of his competitors. Beyond six competitors on a single job, however, individual characteristics are apt to be obscured, and equally satisfactory results can usually be achieved by assuming all competitors to have the characteristics of the typical or average competitor.

The variable markup based on the existing competitive situation is likely to show a marked effect both on the contractor's sales volume and on his profits. Sales volume will decrease as the markup increases; profits, meanwhile, will begin at zero, rise to some maximum value, then decline to zero again as the markup increases.

Table 50 compares the outcome (in terms of gross profits) brought about by using a variable markup with what the actual outcome would have been had various levels of fixed markups been applied by the same contractor on the same jobs described in Tables 48 and 49.

Table 50 shows that, had the contractor bid all 36 jobs at a price only one percent above his estimated direct cost, he would have received 22 jobs, netting $305,000. By increasing his markup to 10 percent, he could have net $888,000 on only nine jobs. But by applying a variable markup—using the theoretically derived optimum bids—his net was actually $1,014,000, realized on 16 jobs.

The amount to be gained by using a variable markup in this case, then, is *at least* 14 percent higher than could have been achieved by *any* fixed level of markups.

This example, by no means unique, certainly presents a substantial argument in favor of a variable markup, even if the markup is based on nothing more than the number of competitors to be encountered on the project.

TABLE 50. Gross Profits Resulting from Various Markups ($1000)

Job Number	Optimum	Markup (%)													
		1	2	3	4	5	6	7	8	9	10	11	12	13	14
1	53	10	19	29	38	48	58	67	77	86	96	106	115	0	0
2	0	0	0	0	0	0	0	0	0	0	0	0	0	0	0
3	0	0	0	0	0	0	0	0	0	0	0	0	0	0	0
4	73	7	15	22	29	36	44	51	58	66	73	80	87	95	0
5	0	0	0	0	0	0	0	0	0	0	0	0	0	0	0
6	0	0	0	0	0	0	0	0	0	0	0	0	0	0	0
7	16	4	8	12	15	0	0	0	0	0	0	0	0	0	0
8	0	3	0	0	0	0	0	0	0	0	0	0	0	0	0
9	230	30	59	89	118	148	177	207	236	266	295	325	354	384	0
10	25	5	9	14	18	23	28	32	37	42	0	0	0	0	0
11	0	22	43	65	0	0	0	0	0	0	0	0	0	0	0
12	0	0	0	0	0	0	0	0	0	0	0	0	0	0	0
13	5	1	2	3	4	5	6	7	7	8	9	10	11	12	13
14	0	0	0	0	0	0	0	0	0	0	0	0	0	0	0
15	0	2	3	0	0	0	0	0	0	0	0	0	0	0	0
16	0	17	34	0	0	0	0	0	0	0	0	0	0	0	0
17	108	36	72	108	0	0	0	0	0	0	0	0	0	0	0
18	0	0	0	0	0	0	0	0	0	0	0	0	0	0	0
19	0	11	22	33	43	0	0	0	0	0	0	0	0	0	0
20	62	15	30	45	60	75	90	105	120	135	150	165	0	0	0
21	0	0	0	0	0	0	0	0	0	0	0	0	0	0	0
22	20	5	10	15	20	25	30	35	40	45	0	0	0	0	0
23	0	0	0	0	0	0	0	0	0	0	0	0	0	0	0
24	23	4	8	12	17	21	25	29	33	37	41	0	0	0	0
25	63	19	38	57	0	0	0	0	0	0	0	0	0	0	0
26	69	15	29	44	59	73	88	0	0	0	0	0	0	0	0
27	0	0	0	0	0	0	0	0	0	0	0	0	0	0	0
28	0	0	0	0	0	0	0	0	0	0	0	0	0	0	0
29	0	41	81	0	0	0	0	0	0	0	0	0	0	0	0
30	0	0	0	0	0	0	0	0	0	0	0	0	0	0	0
31	56	17	34	51	67	84	101	118	135	152	168	0	0	0	0
32	0	0	0	0	0	0	0	0	0	0	0	0	0	0	0
33	15	5	10	15	20	25	30	35	40	45	50	55	60	0	0
34	191	35	69	104	139	174	0	0	0	0	0	0	0	0	0
35	0	0	0	0	0	0	0	0	0	0	0	0	0	0	0
36	5	1	1	2	3	3	4	4	5	6	6	7	8	8	9
Total	1014	305	596	720	650	740	681	690	788	888	888	748	635	499	22

Effect on Sales Volume

Figure 86 shows the effect of various level of fixed markups on the contractor's annual volume. The lower line in Figure 86 represents the estimated cost of the jobs received; the upper line denotes the total sales volume, found by adding the appropriate markup to the estimated direct costs. At a zero percent markup— in other words, by bidding all jobs at cost—the contractor would be low bidder on work costing about $30 million. Raising the markup to 2 percent would have little effect on the amount of work obtained, but an increase to 3 percent would reduce the amount of work received to about $25 million. A further increase in the markup to 4 percent would eliminate another $8 million worth of work. Additional increases in the percentage markup would show less drastic effects, but the volume would gradually decrease until a 14 percent markup was reached, at which point there would be no volume.

Effect on Profits

At zero markup there would naturally be zero profit on the $30 million volume; likewise, at a 14 percent markup there would also be zero profit, this time on zero volume. In between these two unpleasant extremes, the profits can

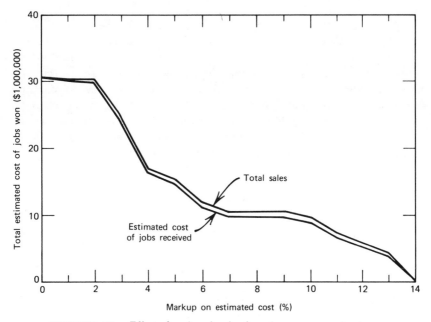

FIGURE 86.   Effect of various levels of markups on annual volume.

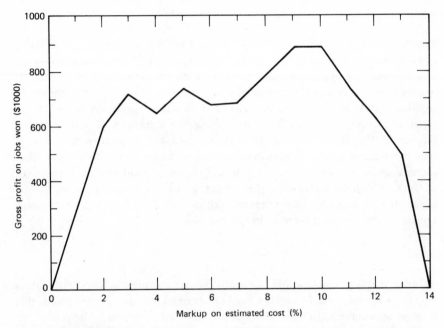

FIGURE 87.   Effect of various levels of markups on annual gross profit.

be expected to rise from zero to some maximum value, then decline until zero is again reached. Figure 87 shows how the profits would have been affected by this contractor's markups. The maximum value possible with a fixed markup would have been realized at a markup in the 9 to 10 percent range, representing work costing $9 to 10 million. This contractor, then, could have realized substantially greater profits by taking a 10 percent markup on $9 million worth of work than he could have earned by taking either a lower markup on a higher volume (5 percent on $15 million, for example) or by taking a higher markup on a lower volume (such as 12 percent on $5 million).

The Success of Strategy

Table 51 summarizes what the actual results of the alternative strategies available to this contractor would have been, in terms of the number of jobs won, the cost of jobs won, the resulting total sales volumes, and the corresponding gross profits.

These results must be considered in view of the contractor's objectives. Using the theoretically correct markups, while resulting in the highest possible gross profits, might not necessarily be the best course to follow in every case. The

$1,014,000 in gross profits attained by using the optimum markups represents an average markup of 4.9 percent on estimated direct job costs.

If no substantial increase in overhead or capital costs were required to support the $22 million volume associated with the optimum bids, over the costs of operating at, say, half that total volume, then the use of the optimum bids is well justified—so long as the contractor's objective is to maximize profits.

If, however, additional costs are likely to be incurred in handling the larger volume, then some intermediate solution is warranted. Probably the most satisfactory solution would be to set a minimum markup, based on the firm's operating and financial requirements and objectives and to look for jobs offering at least that minimum potential.

In the example, had no jobs been bid on which the optimum bid was less than 5 percent, only 16 of the 36 jobs listed in Table 49 would have been bid at all. Of these 16, the contractor would have been low bidder on 8 (jobs 1, 4, 9, 10, 13, 24, 34, and 36). On these 8, gross profits would have been $605,000, on jobs costing $9,147,000, resulting in a total sales volume of $9,752,000. This profit represents an average markup of 6.6 percent on estimated costs.

This substantial level of profits, then, could have been achieved by bidding only 16 jobs—all of which were easily identifiable for some time prior to the letting. And, by devoting more attention to identifying other jobs (not included

TABLE 51. Summary of Results of Using Various Markups

Markup (%)	Number of Jobs Won	Estimated Cost of Jobs Won	Gross Profit on Jobs Won	Total Sales Volume
Optimum	16	$20,701	$1,014	$21,715
0	23	30,618	0	30,618
1	22	30,168	305	30,473
2	21	29,869	596	30,465
3	18	23,938	720	24,658
4	15	16,276	650	16,926
5	13	14,803	740	15,543
6	12	11,331	681	12,012
7	11	9,866	690	10,556
8	11	9,866	788	10,654
9	11	9,866	888	10,754
10	9	8,905	888	9,793
11	7	6,807	748	7,555
12	6	5,303	635	5,938
13	4	3,838	499	4,337
14	2	157	22	179

in the original 36) offering potentially greater profits, the results would probably have been even better.

Using a well thought out competitive bidding strategy in conjunction with management's own goals and requirements cannot help but improve the company's position, regardless of what the goals and requirements are.

Management Review and Revision

A competitive bidding strategy cannot just be implemented and forgotten, but must be continually reviewed by top management to insure the strategy's compatibility with the firm's objectives.

In the strategy fails to accomplish the firm's objectives, then the strategy should be carefully reviewed. If it still fails to meet the firm's requirements, then the objectives should be reviewed. One or the other must be changed.

In theory at least, the competitive bidding strategy should change as each additional bid of competitive information becomes available. In practice, this would mean developing a new strategy each time another job is let on which the competitor's bids become known. Such continual revision is neither necessary nor desirable.

The strategy should be revised periodically, however. The elapsed time between revisions will depend largely on the amount of new data that becomes available. For most contractors, annual or semiannual revisions should be adequate, unless the competitive situation becomes significantly altered.

A complete revision of a competitive bidding strategy will require the addition of new data to existing data, thus changing the frequency distribution of competitors' bids, and subsequently leading to the construction of new probability curves.

In most cases, one, or at most two, years of competitive data are sufficient. As the data from the most recent year's operations are added, the oldest year's data can be dropped. This practice keeps the strategy current.

Effect of Strategic Bidding on Industry Profits

One of the main points that rapidly becomes apparent in developing and applying a competitive bidding strategy is the need for selectivity in bidding. By picking jobs with the greatest profit potential and concentrating on these jobs, the idea of "volume for the sake of volume" will become distasteful to all thinking contractors. The effect in the long run will be to make the contracting industry profit-oriented, rather than, as is largely the present case, volume-oriented.

As has been shown both in theory and in practice, the number of bidders per job has a direct effect on the profits realized by the low bidder. The numerous contractors who feel obliged to bid every job that is offered are chiefly responsible for driving away the industry's profits. Figure 88 shows the general relationship between the number of bidders on a job and the amount of profit in the job for the successful bidder.

From the relationships between the profit in a job and the number of bidders for the job, it is apparent that reducing the number of bidders on each job will result in increased profits for everyone.

For example, assume that there are ten contractors competing for 100 jobs, each job having an estimated direct cost of $100,000.

If each of the 10 contractors bid all the 100 jobs, the average job will, at best, go for approximately 2.4 percent above cost. Total profit on the $10,000.000 worth of work will amount to $240,000, or $24,000 per contractor.

But if each contractor bid only 50 of the jobs, then there would be an average of only five bidders per job. The average net should be approximately 3.3 percent per job, amounting to $330,000 total, or $33,000 per contractor. This is in spite of the fact that each contractor is bidding only half as many jobs as before, and the total amount of available work remains unchanged.

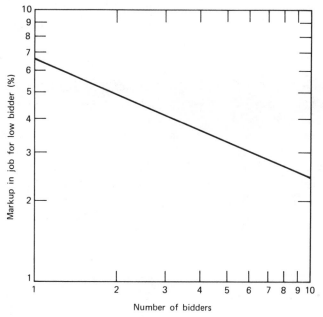

FIGURE 88. General relationship between markups and number of bidders.

The intensity of competition—the main cause of low profit margins—is not simply a function of the amount of work available and the number of firms in the business; it is just as much a function of the number of bids submitted by each contractor. There can be tougher competition, and lower profits, with half as many contractors in the business competing for the same volume, if each contractor insists on bidding twice as many jobs. Profits depend more on the number of bidders per job than on any other single factor.

More bidders per job means lower profits to the low bidder; lower profits on each job means that a higher volume must be obtained to break even; a higher volume requires that more jobs be bid; and bidding more jobs means that profit margins will become even lower. This, then, is truly a vicious circle.

The contractors who apparently expect to reap large profits on low markups—and there are many of them—must have much the same business philosophy as the Japanese automobile dealer who hung the following sign in front of his Tokyo used-car lot:

We sell new used cars for lowest prices anywhere in Japan. We buy old used cars for highest prices anywhere in Japan. How we stay in business? We lucky!

Fortunate indeed is the contractor who can be so lucky.

Summary

A bidding strategy can have several different objectives, and the strategy's effectiveness can be measured in terms of its ability to achieve the desired objectives better than any other predetermined plan. Theoretically derived bidding strategies have proven effective in a number of cases in achieving the maximum profits possible under existing competitive conditions; for whatever the level of markups that has been employed by the contractor, this level has probably failed to achieve profits as high as could have been obtained by some other strategy. A variable markup, based on the amount of competition expected, will almost invariably result in higher profits than any constant markup. Regardless of the strategy employed, it should be continually monitored by management, both to make sure the strategy is accomplishing its objectives and to make sure that the firm's objectives are reasonable in view of the existing competitive situation. The overall effect of strategic bidding will be to raise the general level of industry profits by making individual contractors profit-conscious rather than volume conscious and by decreasing the intensity of competition through increased emphasis on the need for more selective bidding.

References

Benjamin, Neal B. H. *Competitive Bidding for Building Construction Contracts.* Technical Report No. 106, Department of Civil Engineering, Stanford University, June 1969.

Benjamin, Neal B. H. *Competitive Bidding: the Probability of Winning.* Journal of Construction Division, ASCE, September 1972.

Casey, B. J. and L. R. Shaffer. *An Evaluation of Some Competitive Bid Strategy Models for Contractors.* Report No. 4, Department of Civil Engineering, University of Illinois.

Churchman, C. W., R. L. Ackoff, and E. L. Arnoff. *Introduction to Operations Research.* John Wiley & Sons, 1957.

Edelman, F. The Art and Science of Competitive Bidding. *Harvard Business Review,* July–August 1965.

Friedman, L. *A Competitive Bidding Strategy.* Operations Research, No. 4, 1956.

Gates, Marvin. *Aspects of Competitive Bidding.* Connecticut Society of Civil Engineers, 1959.

Gates, Marvin. *Statistical and Economic Analysis of a Bidding Trend.* Journal of Construction Division, ASCE, Paper No. 3210, November 1960.

Gates, Marvin. *Bidding Strategies and Probabilities.* Journal of Construction Division, ASCE, Paper No. 5159, March 1967.

Gates, Marvin. *Bidding Contingencies and Probabilities.* Journal of Construction Division, ASCE, Paper No. 8524, November 1971.

Mayer, R. J. Jr., R. M. Stark, and W. Fitzgerald. *Unbalanced Bidding Models— Applications.* University of Delaware Technical Report, March 1969.

Morin, T. L. and R. H. Clough. *Opbid—Competitive Bidding Strategy Model.* Journal of Construction Division, ASCE, Paper No. 6690, June 1970.

Park, W. R. How Low to Bid to Get Both Job and Profit. *Engineering News-Record,* April 19, 1962.

Park, W. R. Less Bidding for Bigger Profits. *Engineering News-Record,* February 14, 1963.

Park, W. R. Bidders and Job Size Determine Your Optimum Markup. *Engineering News-Record*, June 20, 1968.

Park, W. R. How Much to Make to Cover Costs. *Engineering News-Record*, December 19, 1963.

Park, W. R. Profit Optimization Through Strategic Bidding. *AACE Bulletin*, December 1964.

Rubey, H. and Walker W. Milner. *Construction and Professional Management*. Macmillan, 1966.

Shaffer, L. R. Competitive Strategy Models for the Construction Industry. *International Journal of Computer Mathematics*, No. 1, 1965.

Shaffer, L. R. and Terry Micheau. *Bidding With Competitive Strategy Models*. Journal of Construction Division, ASCE, Paper No. 8008, March 1971.

Stark, R. M. *Unbalanced Bidding Models*. University of Delaware Technical Report, 1966.

Stark, R. M. *Unbalanced Bidding Models—Theory*. Journal of Construction Division, ASCE, Paper No. 6174, October 1968.

Stark, R. M. and R. H. Mayer, Jr. Multi-Contract and Unbalanced Bidding Models. *Bulletin of the Operations Research Society of America*, No. 16, 1968.

Stark, R. M. and R. H. Mayer, Jr. *Static Models for Closed Competitive Bidding*. University of Delaware Technical Report, 1969.

Index